About the Author

The author has spent over 30 years studying the sciences, as well as global politics, finance and religion. His main goal has been the study of nineteenth century Aether, and how it might hold the key to solving many of the problems facing modern science. From that basis, the author has attempted to find a common thread linking international finance and politics with the sciences, and how all three are being manipulated for the purpose of setting up a New World Order.

SECRETS OF THE UNIVERSE REVEALED

The Tesla Legacy

TONY CLARK

YOUR BOOKS

First edition published 2019
Publisher: Anthony William Clark
Printer: Your Books
18 Cashew Street, Grenada North
Wellington
NEW ZEALAND

www.yourbooks.co.nz

ISBN 978-0-473-47150-7

Typeset in Adobe Garamond Pro
Cover concept: Alissa and Lance Harper
Cover artwork: Justine Woosnam
Book design: Bozidar Jokanovic

Contents

Preface 9

Nikola Tesla 13

War Of The Currents 21

Discovery Of Darkness 33

Resonant Power 41

Colorado Springs 59

Wardenclyffe 71

Visions Of Utopia 95

A House Divided 119

Rosetta 135

Aetheric Warfare 167

East Meets West 205

The City 231

The Central Bankers 245

World Government, By Consent 263

World Government, By Conquest 279

Beyond The Nation-State 305

The Key 317

Glossary 325

Acknowledgments

My greatest inspiration for doing this book is the Bible. So first and foremost, I give all glory to my Saviour, the Lord Jesus Christ.

Over the last 30 odd years, there have been many people who have inspired me to change direction in a particular area of research. Many Christian, as well as secular authors, have helped lay the foundations for what I believe when it comes to finance, politics and the sciences.

Below are some of the people I want to thank for helping me along the way.

Leona Clark (Co-Editor)
Fern Lowes (Typing and Formatting)
Justine Woosnam (Artwork and Formatting)
Alissa and Lance Harper (Designed Front Cover)
Nelson Clark
Christine and David Willis
Suzanne Robson
Gary Clark
Abel Clark
Mark Clark
Terry Staunton
Tewhiu Isaac
Rodger and Rene Downie and family
Hazel and Ivan Radford and family
Peter and Alison Radford
Adam Parker
Lynn and Alan Harper
Elisabeth and Michael Brown
Peter and Leonie Claydon

PREFACE

"If you never know a shot has been fired, you may not even know you're at war, until you've already lost."

–Author Unknown

SINCE CHILDHOOD, I have always had a deep fascination for the sciences, especially when it comes to astronomy and physics. As a young pupil, I learnt a lot from teachers and books, but there were always more questions than they had answers for. On one occasion, my teacher was instructing the class on how the sun used gravity to hold the earth in place, after pondering his statement for a while, I came away dissatisfied with the argument he presented. If gravity was an attracting force, then why hadn't the earth sunk into the fiery depths of the sun eons ago. Even at a young age, doubt arose about how scientists were interpreting the information they were gathering.

It was in my late teens, when I discovered two books that would forever change my perception of how the universe worked. They spoke of a long-lost concept known as the Aether, and that it permeated every part of the universe. Feverishly studying the theory, it wasn't long before I realised why it wasn't accepted by the scientific community, despite recent discoveries that might actually support its existence.

For the next fifteen years, my interest in the Aether never waned, particularly when I discovered it could be used to manipulate the weather. But I had to admit, many questions remained about why the theory was being deliberately pushed aside by scientists, just maybe I had deceived myself into believing something was there when it wasn't.

However, doubt would disappear for good, when a few of us got to witness a very unusual event in the sky.

In early 2006, while living in the city of Christchurch, New Zealand, the late summer weather had been dry and settled, with no forecast of rain or cold temperatures in sight. Then out of the blue, four cold grey days descended on us, it was as if someone had grabbed the Roaring Forties off the southern ocean and placed it over this part of the country. It didn't snow, but it was like the freezer door to Antarctica had been opened. I even remember saying to people how strange this weather was, compared to how it had been throughout the summer months.

Then after the fourth day, the cold spell broke and barring a bit of high cloud, the sky was blue again. That morning, as we sat outside the house enjoying a cup of coffee, we looked up and there directly above us was a cloud formation that we had never seen before. Rushing inside, I grabbed a small camera, then quickly made my way back out to take some photos of the *grate* like pattern.

Even though it took less than sixty seconds to get the camera, the high-altitude winds had begun to dissipate the perfectly aligned set of clouds. Slightly disappointed the clouds had lost their form, we sat at the small table in absolute wonderment at what had just been observed, but little did we know the best was yet to come.

As I occasionally scanned the sky, a second cloud formation appeared directly south-east of the original. The second one was shaped like a *fan*, (Figure 18, page 164), and was caught on film just after it appeared. Timing was important, because within twenty seconds of appearing the high-altitude winds had made it unrecognisable.

With camera in hand, I paced up and down the yard in the hope of maybe seeing something else. I was not to be disappointed, within ten minutes of the *fan* like formation, a third one appeared directly in line (south-east) with the other two. Looking like, a *donut* cut in half (Figure 19, page 164), the last photograph shows just how big the total area of the three formations was.

From that moment on, the subject of Aether physics, and its use as a new type of weapon system had finally become very real to me. In my own mind, I could now accept that Aether theory wasn't just correct, but was engineerable, now all I wanted to know is, who controlled its use. Was it owned and run by officials from various countries, or was it under the control of some kind of clandestine group?

Many documented cases exist of people who have made discoveries into Aether technology, but because of outside interference, they have been forced to quietly withdraw from public view.

A few of the more notable figures include T. Townsend

Brown, T. Henry Morey and Royal Raymond Rife. But one inventor stands out from the rest, he is considered the pioneer of Aetheric research, and his name is Nikola Tesla. Some say it was the development of his special transformer in the 1890s, which allowed the all-pervading field to be harnessed. If that is correct, then someone is suppressing the truth about this revolutionary technology and its potential as a weapon. And that can only mean one thing, the nations of the world are up against a weapon of unimaginable power, and if they don't wake up, then those who control it will be free to drive us into an Orwellian global order of their making.

CHAPTER ONE

NIKOLA TESLA

"Truth is stranger than fiction, but it is because Fiction is obliged to stick to possibilities; Truth isn't."

–Mark Twain, 1897

EARLY YEARS

MILUTIN TESLA'S PHILOSOPHICAL nature and infallible memory made him a very astute writer. This Orthodox priest of Serbian descent wrote many editorials for the newspapers concerning important social issues of the day. The intellectual elite and dignitaries were so impressed with his emotional articles, that awards were presented for his thought-provoking work.

Milutin also had a very dry sense of humour, telling a friend who accidentally hung his coat over the side of the carriage and onto the wheel, "Pull your coat in, you are ruining my tyre." On the other hand, because he held lengthy conversations with himself, where the tone of his voice would change between sentences, some people might have mistakenly come to the conclusion he was insane.

In 1847 Milutin married Djuka (pronounced Juka) Mandic, the daughter of a prominent Serbian family. Djuka was extremely gifted, and even though she couldn't read or write, she was able to phonetically memorize great epic Serbian poems, as well as long tracts of the Bible. She was a woman of rare skill and hard work, labouring from dawn till late in the night making most of the family's clothes, and furnishings by hand. She also built many labour saving devices for the local women; her intention wasn't to make money but to build inventions that made them and their family's lives more comfortable. Later in life, Nikola Tesla commented that even past the age of sixty her fingers were still nimble enough to tie three knots in an eyelash.

Later, the Tesla family moved to the small town of Smiljan, where their father Milutin was able to minister to a larger congregation. This was then followed by a fifty-mile move to the Lika Valley in the district of the same name.

On the 10th of July 1856, a legend states that at the stroke of midnight, during an electrical storm, Nikola was born. He was the second youngest of five children, and while he grew up doing all the things that normal boys do, he also developed a fascination for his mother and older brother Dane's elaborate inventions. By the age of six, his inventive streak had become quite prolific. One invention he built was a self-rotating propeller, that involved sticking sixteen live maybugs to a wheel, so when they flapped their wings the wheel started rotating. But there was a downside to his revolutionary device, and that was it came to an untimely end when a local lad uninceremoniously ate the maybugs, revolting the young inventor beyond words. His far-fetched attempt at building some kind of generator might seem ludicrous to most observers, but it was just a sign of what was to come.

It was his inventive prowess in conjunction with his eidetic memory (photographic memory, and ability to be able to build, test and modify a device in his mind) that allowed his creativity to run wild.

Nikola Tesla's eidetic memory was extremely powerful. It is said, how sometimes he would need the help of his sisters to differentiate between reality and the images permeating in his head. This mental ability allowed him to do more than just memorize huge amounts of poetry, or learn several languages, it also enabled him to design machines with his mind.

Then without warning, the Tesla family's faith was tested when an event took place that shook the family to its core. Dane, Nikola's older brother, died from a horse riding accident in 1863. It devastated his parents to such a degree that they would never completely come to terms with his passing. Nikola, who was extremely talented like his brother, would only remind his parents of the older sibling. This caused an uneasy atmosphere to develop in the home, which made the young lad withdraw into himself. Even as an old man Tesla would recall the painful scars his older brother's death inflicted.

Later on, when the Tesla family moved to the city of Gospic, Nikola found it hard to fit into the urban way of life; his love for nature and the country allowed his creativity to flourish, but the city seemed to stifle his inventive streak. Worse still, on one occasion while employed as the bell ringing boy in his father's church,

Nikola managed to tear the beautiful dress of one of the most influential women in the city; making the young lad even more of a pariah than he was before.

But all of that changed when the city fire company obtained a new pump to help fight fires; it was a far cry from the days of passing buckets from one person to the next. The city of Gospic and its fire company were so proud of their new apparatus, that they wanted nothing more than to show it off to its citizens. When the moment arrives to show the public, the pump is finally engaged, but to their horror, nothing comes out of the hose, and even after many adjustments they still couldn't get it to work. Grabbing one of the officials, Nikola exclaims "I know what to do", so he takes off his outer clothes and jumps into the river. He then proceeds to unkink the hose which feeds the pump, and in doing so the water bursts forth, bringing cheers from the delighted crowd. Tesla is the hero of the day, and to say the least he enjoys every second of his newfound adoration.

With his confidence high, the young Tesla ploughed through his studies, putting four years into three, but there was still one obstacle standing in the way of his dream of becoming an engineer—his father Milutin. The Orthodox priest wanted his son to follow in his footsteps by going to seminary, but to the surprise of the boy, an event would occur that would change his father's mind for good.

During this time an outbreak of cholera hit the area, causing the young student to fall severely sick, and at one point even coming close to death. Even with his sickness, Nikola had a plan. He would say to his father "Perhaps I may get well if you let me study engineering?" with his father replying "You shall go to the best technical institution in the world." When he eventually got over his sickness, nothing could hold the budding engineer from achieving his goals. And after a short stint in the military, which he later recalled as being utterly barbaric, Tesla was inducted into the Polytechnic School at Graz, Austria. Throwing himself mercilessly into his studies, it wasn't long before the scorecards showed A+'s across the board, showing just how easy higher learning was for the young Serbian.

On one occasion while in class, a confrontation occurred between Tesla and his physics teacher, Professor Poeschl. During the lecture, a Direct-Current (DC) Gramme dynamo was delivered to the professor's class; this was before the days of Alternating-Current (AC). When Tesla stated that a particular device on the motor, called the commutator wasn't necessary and that the more powerful alternating, (back and forth motion), current could be harnessed, the professor proceeded to belittle the student in front of his peers, even pulling the commutator out and saying, "See! It doesn't work." This drove Tesla into an all-consuming battle to prove the professor wrong. But because his quest began to consume his every thought, he started to slip in his studies, eventually failing to graduate from Graz Polytechnic.

Going cap in hand to his family, he then enrolled for a term at the Charles-Ferdinand branch of the University of Prague, and it is there that Tesla later states how he made a significant leap forward in his understanding of the AC process.

Tesla's next step was to join the fledgling electrical industry, so he could learn as much as possible about electricity, and how to harness the far more efficient back and forth motion of Alternating-Current.

AC/DC

With his higher learning coming to an end, Tesla eventually moved to Budapest, Hungary, with his friend Anthony Szigeti. They gained employment with the Budapest Telephone Exchange, giving them insight into the latest technologies available. Some of the equipment they were using was the handiwork of one Thomas Alva Edison, a man who would later become the other half of one of the greatest industrial rivalries the world has ever seen. During this time, the two engineers spent every waking moment combing the literature and modifying equipment, trying to eliminate the commutator from the DC generator. But because of his non-stop obsession, Tesla's emotional and physical health began to suffer, and in the end, he almost succumbed to a total breakdown. The next few months were critical in bringing Tesla back to health, and it was Szigeti's determination to help his friend recover, that would lead to the discovery of how to do away with the commutator.

One day while Szigeti was trying to walk his friend back to health in a local park, Tesla gazed upon the setting sun while quoting Goethe's *Faust*:

> *"See how the setting sun, with ruddy glow. The green embosomed hamlet fires. He sinks and fades; the day is lived and gone. He hastens forth new scones of life to waken. O for a wing to lift and bear me on, and on to where his last rays beckon."*

Tesla declares it was at this moment that it all became clear on how the new system would operate. Proceeding to poke a stick in the sand he feverishly drew his new-found revelation. What Tesla had discovered was the Rotating Magnetic Field. The reason Alternating-Current (AC) was far superior to Direct-Current (DC), was the former could use transformers to raise and lower the voltages (pressure), thus sending electricity for hundreds or thousands of miles, while the latter could not.

To use an analogy, imagine that AC is like a huge water pipe that allows large amounts of water (electricity) to flow through it, and when you want to draw some, just put a valve (transformer) onto the pipe and bleed some of the water off for personal use. Then there is DC which, because of its design, only allows for a small pipe from start to finish, and so the friction of the pipe retards the flow until there is no pressure left.

When it comes to the DC motor the electricity is produced mechanically through the commutator, but for AC it is produced electronically. This makes the AC motor less complicated, and thus cheaper to make. The only downside with AC, is if you are one of the unfortunate souls who accidentally touches a wire—which leaves one with an indelible memory, to say the least.

Through his new found insight into the principles of AC induction, Tesla would eventually go on to develop not just the induction-motor, but also introduce the other components needed to make the AC system work, like the transformer developed for Westinghouse by William Stanley.

In 1884 Nikola Tesla embarked on the next leg of his AC endeavour; he knew if he was to make his discovery a reality he would have to go to the United States, a nation that was in the throes of a full-blown industrial revolution, and eager to embrace new and innovative ideas. Tesla finally arrived safe and sound, despite being robbed of his possessions, and a ship-born mutiny which almost saw him being thrown overboard.

Unfazed by his travelling misadventures, the inventor threw himself into the New York way of life and wasted no time in producing a letter of introduction to his new employer from his old boss in Europe. Tesla wasn't just meeting his idol; he was going to work alongside the man who would become known as the Wizard of Menlo Park, Thomas Alva Edison.

As the story goes, after settling into his job, Tesla submits his plans of Alternating-Current to Edison. He politely explains how it is superior to Edison's own Direct-Current, which unsurprisingly falls flat with the Menlo magician. One has to remember that long before the young upstart had stepped onto American shores, Edison had been considered an inventive genius. His own patents included direct current, the phonograph, and the incandescent light bulb, to name a few. No one could deny the place this man held in American industrial folklore. But the simple fact was, Edison had spent so much time and money on DC that he was not going to drop it no matter how superior Tesla thought AC was.

CHAPTER TWO

WAR OF THE CURRENTS

"I was fascinated by a description of Niagara Falls I had perused, and pictured in my imagination a big wheel run by the Falls. I told my uncle that I would go to America and carry out this scheme. Thirty years later I saw my ideas carried out at Niagara and marvelled at the unfathomable mystery of the mind."

–Nikola Tesla

AFTER SOME ADVICE from his old boss in Europe, Tesla decided to help Edison improve the DC system. Of course, it was aided by the incentive of a $50,000 bonus upon completion. This prompted the young engineer to throw himself fully into the project, working day and night, with back to back shifts on many occasions, until the job was completed. When it came time to collect the money, Edison is said to have remarked "You are still a Parisian. When you become a full-fledged American you will appreciate an American joke." Shattered, Tesla promptly quit and started digging ditches for a living. He wasn't too proud to perform manual labour, even though it was for a pittance compared to what Edison had been paying him. Tesla knew that with his qualifications, it would only be a matter of time before some other financier or industrialist came calling, then he would be back in business.

After a few false starts, Tesla formed a company with a lawyer called Charles F. Peck. The inventor needed his support but this lawyer was no pushover; he wasn't going to spend his hard-earned money on some air head's dream, he wanted proof this guy knew what he was talking about. Desperate, Tesla knew he had to win the investor over, so he had an inspiration. There was an old legend that says when Christopher Columbus was trying to gain royal approval for his adventure of discovery to the New World, he impressed certain dignitaries at a dinner by telling them that he could balance an egg on its end. After many failed attempts by other guests, Columbus proceeded to gently knock the egg, slightly cracking it, and to the amazement of the onlookers it stood on its end, and needless to say, Columbus got his audience with the Queen and the rest they say is history.

Tesla states to Peck that he can do the same—but without cracking the egg—drawing the interest of the lawyer. Tesla then states, "If I can do this will you invest?" The desperate inventor rushes to the local blacksmith and has an egg made of iron and brass, then he constructs an enclosure with a polyphase electric fence around the periphery. Once the egg is placed in the centre and the current is turned on, it begins to spin, then as it increases in speed it sits up on its end just like Tesla said it would.

Not much more needed to be said. Tesla, Peck and a third partner Alfred S. Brown went on to form the Tesla Electric Company.

It was at this point that many of the downstream devices were developed for the AC system. Tesla was so obsessed with tying up all possible patents that he engineered one, two and three phase, but he also theoretically developed four and six phase current systems as well. Time, however, was running out, so the inventor and his partners moved ahead with only some of the fundamental patents in place. Tesla, Peck and Brown had to start courting big money interests in the hope of using their commercial might to push AC through the mental barrier the electrical fraternity had with Tesla's system. This mental barrier wasn't because of arrogance or greed, it was because they couldn't understand why it would be better than the existing systems, and it didn't help that Tesla was extremely cagey about releasing information before all the pieces were in place, particularly the patents.

Also, early forms of AC already existed but they were hampered by extreme inefficiency, and even though Tesla eventually released his version of AC to the world, some of these same men would try and sue the inventor for patent infringements. But unlike others, Tesla's system was completely different and had the ability to send electricity for hundreds of miles. Despite various setbacks, it was men such as T.C. Martin in his editorials for the *Electrical Engineer* that helped push Tesla's AC system into public consciousness. This exposure was enough to draw the big industrial corporations into taking an interest, and it was a man by the name of George Westinghouse who would eventually court the three entrepreneurs.

Westinghouse, of Pittsburgh, Pennsylvania, was an

inventive genius in his own right. One of his greatest claims to fame was the railroad airbrake. Before his invention, trains had very little braking capacity, which culminated in many railroading disasters. Widely known amongst his employees as a very generous man and positive motivator, the industrialist was also known for having an eye for new technologies and how to commercialise them for the marketplace.

Tesla and his partners eventually agreed to sell the patents to Westinghouse, with some accounts suggesting they sold it to the industrialist for the knockdown price of $500,000, while others said it was for even less. But it didn't matter, because it was the royalties of $2.50 per horsepower, on all alternating current capacity sold, that would have made the three men wealthy beyond their wildest dreams.

After the agreement, Tesla went to Pittsburgh to help in the development of his induction motor, and so all the pieces, including the big financial benefactor, were in place. It has to be said, even though there were some satisfying victories along the way, the fight was far from over. Standing in the way was Tesla's old boss, the one who would vehemently fight him so that his own DC system could be victorious.

Thomas Edison's dream of bringing the DC system to the world was flawed. Its gross inefficiency meant the average city would need several big coal fired power plants to get electricity to the city limits, so it could be imagined how bad the pollution would have been. Also, because of the cost of building expensive power plants, the surrounding countryside and small towns would have had to remain in the dark.

The burgeoning industrial age was screaming out for

a better way to distribute energy; the antiquated steam driven system was already losing pace. But for Tesla, there was a problem, the DC system was already starting to penetrate the market, while AC was, by and large, an unknown technology.

The war for recognition wasn't really between Tesla (AC) and Edison (DC), it was between George Westinghouse (Westinghouse Corporation) and J.P. Morgan (General Electric), the backer of Thomas Edison and the man who later became Tesla's greatest adversary. Edison knew Tesla was a threat, so his propaganda machine went into overdrive, including a plan to show the public just how dangerous AC was. Because AC required a much higher voltage to operate, Edison keenly demonstrated the potential danger by electrocuting small animals and even a fully grown elephant.

The situation only got worse when several cities decided to kill their stray animals in this fashion, a situation that put AC in an extremely bad light with the public. Then when it couldn't get any worse, the use of AC power by H.P. Brown in his invention of the Electric Chair, was the point where the last nail could have been placed in the proverbial AC coffin. This type of execution was even nicknamed "Westinghoused", and it left nothing to the imagination when it came to cruelty. The first unlucky criminal to endure this form of capital punishment was a man by the name of William Kemmler.

To make matters worse, on the day of Kemmler's execution, the procedure was bungled to such a degree, that a savage outcry arose from people across the nation, for all agreed that Kemmler's death was nothing short of torture.

By this time, George Westinghouse had come under intense pressure from his investors as well as the public. They wanted him to put an end to the high voltage AC system Tesla was perfecting for his company, but Tesla must have pleaded with the clever industrialist to stick with it. Whatever was agreed upon it must have worked, because later in life he states.

> *"George Westinghouse was, in my opinion, the only man on the globe who could take my alternating current system under the circumstances then existing and win the battle against prejudice and money power. He was a pioneer of imposing stature and one of the world's noblemen."*
>
> **—Nikola Tesla**

It seemed the alternating 'light' was finally starting to flicker through the dark haze of public opinion, invigorating some into conducting new experiments into the transmission of AC power over longer and longer distances. But it was two up and coming projects by Tesla, that promised to catapult the fledgling electrical industry into the mainstream. One project was the Chicago World's Fair, which was to be electrified for the very first time. The second was the building of a power plant at the world famous Niagara Falls.

Also known as the Columbian Exposition, this immense expo within the city limits of Chicago was seven hundred acres, had sixty thousand exhibitors, and cost twenty-five million dollars. In the end, it would accommodate a staggering twenty eight-million patrons over its six-month life. The massive fair had the ability to show the world what was possible in this new technological age. And for Edison and Tesla, they too knew the importance of what electrifying the exposition would mean. But the cost of such an undertaking was far beyond the means of the two inventors, so the money had to be supplied by Westinghouse and Morgan. But despite the enormous financial outlay, there was no doubt neither camp could let this opportunity slip by; the penalty for not participating would see the loss of millions of dollars already invested, not to mention the billions in lost future earnings.

After many years of arduous work and unflinching dedication, success finally came for Tesla and the Westinghouse consortium. AC was now the winner to power and light the Chicago World's Fair.

The victory was not without problems though, the bulbs needed to light up the Expo were in fact built by Edison, and he wasn't about to let his opposition get a leg up. As luck would have it George Westinghouse had a backup. He already owned his own bulb, and even though it was inferior to Edison's, it would do the job required of it for the life of the Exposition. Now all he had to do was make two hundred and fifty thousand of them, and he had just six months to do it. It would be a massive undertaking to achieve, but achieve it they did—and in the process, made the 1892 Chicago World's Fair a roaring success. For the first time in the history of the Exposition, AC was used to light up the night with nothing more than a flick of a switch.

There is one interesting footnote when it comes to Tesla's various displays at the Chicago World's Fair: not a lot is known about it but it must have left the audience spellbound. In his book *Wizard, The Life and Times of Nikola Tesla* author Marc J. Seifer describes what an unforgettable spectacle it must have been.

> *"The simultaneous operation of numerous balls, pivoted discs and other devices placed in all sorts of positions and at considerable distances from the rotating field. When the currents were turned on and the whole animated with motion, it presented an unforgettable spectacle. Mr Tesla had many vacuum bulbs in which small light metal discs were pivotally arranged on jewels and these would spin anywhere in the hall when the iron ring was energized."*
>
> **–Marc J Seifer**
> **Wizard, The Life and Times of Nikola Tesla**

As patrons observed the exhibit, they must have wondered at the incredible possibilities this wireless system presented, especially if it could do away with wires and transformers all together.

The establishment of a Commission, to oversee the development of the world's first hydro-electric scheme at Niagara Falls, had the embryonic electrical industry clamouring to take part in this major undertaking. Whichever electrical system got the go-ahead to use the mighty cataract to turn its turbines, left no one in any doubt, as to who would win the war.

The Commission, set up to oversee the project, took submissions from various groups. These included mechanical and electrical methods, with six of the eight electrical plans involving Direct-Current (DC).

Professor George Forbes, an adviser overseeing the Commission, made it clear to his colleagues that the only workable system the Niagara Power Company should seriously be considering was the Tesla AC motor, which the professor had rigorously tested at the Westinghouse corporate workshops.

After vigorous negotiations between the interested parties, the Commission realized that the only way forward for the mighty project was to use Westinghouse's revolutionary AC polyphase electrical system. The success of the AC system at Niagara, and for the world, was now beyond any doubt. A quote from the imminent Professor Dr Charles F. Scott, who lectured at Yale University, and was the former President of the American Institute of Electrical Engineers, as well as an engineer for Westinghouse when the company helped Tesla develop the AC system, explains what had been achieved at Niagara Falls.

> *"The simultaneous development of the Niagara project and the Tesla system was a fortuitous coincidence. No adequate method of handling large power was available in 1890; but while the hydraulic tunnel was under construction, the development of polyphaser* [Polyphase] *apparatus justified the official decision of May*

6, 1893, five years and five days after the issuing of Tesla's patents, to use his system. The polyphaser method brought success to the Niagara project; and reciprocally Niagara brought immediate prestige to the new electric system.

Power was delivered in August 1895 to the first customer, the Pittsburgh Reduction Company (now Aluminium Company of America) for producing aluminium by the Hall process, patented in the eventful year 1886 ...

In 1896 transmission from Niagara Falls to Buffalo, 22 miles, was inaugurated. Compare this gigantic and universal system capable of uniting many power sources in a super powerful system, with the multiplicity of Lilliputian "systems" which previously supplied electrical service.

As Mr Adams aptly explained "Formerly the various kinds of current required by different kinds of lamps and motors were generated locally; (D.C. system) by the Niagara Tesla system only one kind of current is generated, to be transmitted to places of use and then changed to the desired form (voltage).

The Niagara demonstration of current for all purposes from large generators led immediately to similar power systems in New York City—for the elevated and street railways and for the subways; for steam electrification; and for the Edison systems, either by operating substations for converting alternating current to direct current or by changing completely to A.C. service... By exchange of patent rights, the General Electric Company obtained license rights under Tesla patents, later made impregnable by nearly a score of court decisions."

The evolution of electric power from the discovery of Faraday in 1831 to the initial great installation of the Tesla polyphaser system in 1896 is undoubtedly the most tremendous event in all engineering history."

–Professor Charles F. Scott
Published in Electrical Engineering,
August, 1943

It is no stretch to claim, that the AC polyphase system was one of the biggest contributors to creating the second industrial revolution. It allowed modern industry to grow far and wide through the use of high voltage transmission lines, which were then transformed down to lower voltages for private users in remote areas.

While it appeared as a complete validation to the superiority of the AC system, and an outright commercial victory for the Westinghouse Corporation, the quote by Professor Scott states how General Electric used the courts to circumvent the absolute primacy Tesla/Westinghouse had over the AC patents. It has to be remembered that the war of the currents cost George Westinghouse a lot of money to win, so it left him incapable of totally controlling the electricity market. It seemed General Electric would share in the spoils after all, and then use the courts to cement its position.

To this day the winner of the war of the currents is

officially known as Tesla/Westinghouse and not Edison/ General Electric (J.P. Morgan), but was it?

While Tesla and Westinghouse held the patents, and got the larger share of the contracts (e.g. generators, switch-gear, and auxiliary equipment), General Electric got the transformers, transmission lines, and other equipment, meaning Tesla and George Westinghouse held all the cards, yet somehow J.P. Morgan ended up with the joker.

As for Tesla himself, the average person credits Edison with the invention of long distance electrical transmission, yet they are left none the wiser concerning its true inventor. The individual has to ask themselves, was this historical inaccuracy sheer chance, or was it by design? After much investigation it leaves one without any doubt that the latter scenario must indeed be true.

CHAPTER THREE

DISCOVERY OF DARKNESS

"According to wave radio principles, these systems should never operate. Those who attempt to analyse Tesla impulse transmitters from the radio perspective walk away either baffled or critical. One cannot rationalize Tesla patent texts when viewed in a conventional vein."

–Gerry Vassilatos
Secrets of Cold War Technology-
Project Haarp and Beyond,
2000

THE PHENOMENAL SUCCESS of Tesla's AC system brought him global fame, as well as accolades and invitations from around the world. The prolific inventor was now the man about town, dining out at his favourite restaurant Delmonico's, with New York's elite, such as John Jacob Astor, William K. Vanderbilt and Mark Twain.

Enjoying immense notoriety, the Serbian was secure in the knowledge his royalties would allow him to pursue some of his more mysterious electrical theories, concepts not associated with the AC system, but of a nature so esoteric they would bring him up against the entire scientific community.

It is not exactly clear when Tesla obtained this new discovery, or first formulated the concepts that would lead to his most profound inventions. Some say he obtained it from an old hermit (Nathan Stubblefield), in the woods of Kentucky, while others believe it was when he did his experiments into electromagnetic waves while trying to prove/disprove Hertz's theories; yet others believe Tesla discovered it while working on Edison's DC transmission system shortly before their feud. Whether his breakthrough is from one of these events or at some other time, one thing is for sure—the implications of his discovery were beyond anything in man's wildest imaginations.

Through the use of a bank of capacitors, Tesla started a series of experiments, where he sent high voltage and high frequency DC bursts down a wire. Similar to what is known as a voltage spike, it has the ability to destroy electronics with ease. But what Tesla had also discovered, was that these powerful electrical bursts produced an unusual side effect; there appeared to be a secondary type of shock-wave travelling off the surface of the wire, at right angles. To use an analogy, the secondary shock-wave appears just like the legendary sonic boom, when a jet flies beyond the speed of sound. The shock-waves were so powerful that they had the ability to vaporise the wire, or even send them out into the air, stinging the startled inventor.

Tesla called this electrical effect "Supercharging", and he calculated that it was several times larger than the voltage the capacitor supplied. The only question was, where was the extra energy coming from? Through further experimentation, he discovered that the effect took place as soon as the current was switched on. So it appeared the current was 'bunching up' just as the switch was thrown, then after a moment it would even out, making the strange effect disappear.

It didn't take long for the inventor to realize that he could make the mysterious energy come off the wire by controlling the voltage bursts, and how it could also be amplified for producing physical effects upon objects at large distances.

Even after building new and improved copper and glass shielding to protect himself, he was still unable to stop the shock-waves from hitting his body, proving that these highly penetrative waves were not electrostatic in nature. A quote by Gerry Vassilatos, in his book *Secrets of Cold War Technology-Project Haarp and Beyond*, states what Tesla believed about the energy involved.

> *"If this was electrostatic energy, it was more intense and more penetrating than any electrostatic field he had ever observed... Tesla began to believe that he had discovered a new electrical force, not simply a treatment of an existing force. It is for this reason that he often described the effect as "electrodynamic" or "more electrostatic"."*

> *–Gerry Vassilatos*
> *Secrets of Cold War Technology-*
> *Project Haarp and Beyond,*
> *2000*

The inventor realized that he had to handle the shock-waves and their voltages with extreme care, because they could have killed him with relative ease. At that time, it wasn't unheard of for some of those working on the large commercial electrical systems to be killed. When currents were switched on, they would 'bunch up', and cause electrostatic (lightning) discharges to leap off the circuit in search of the nearest grounding. Not until they redesigned the high voltage DC system, did they manage to engineer the effect out.

Tesla wasn't the only person to experience these mysterious effects. In 1872, an account mentioned in the journal, *Scientific American*, detailed how physicist Elihu Thomson observed "strange action at a distance" effects, when connecting a Ruhmkorrf Coil, to the metal water mains. Thomson and colleague Edwin Houston discovered that the brass door knob on the wooden door of their lab, gave off a static shock when the coil was turned on. But when the two physicists went throughout the university buildings, they discovered, that other insulated metal objects also exhibited similar effects as the door knob. What is not known is whether Thomson ever discovered the underlying physical principles governing this phenomenon.

Conversely, Tesla was quite confident he wasn't dealing with high frequency-alternating currents, like the ones produced during his AC experiments. His new discovery lent more towards *Radiant Electricity*, or power that travelled through 'free space'. It is an energy made up of *Impulse Currents* (think, pearls on a string), not the normal electromagnetic waves (think, swaying fish tail) that we are all familiar with. Through observational evidence, he knew his *impulse currents* went straight through objects of high insulation, causing real effects on the other side. And in Elihu

Thomson's case those effects went straight through walls, rooms, and buildings throughout the campus, causing resonant effects on any object tuned to the frequency of the coil.

Tesla now knew the energy he was dealing with could not have been carried by the air particles around him; it had to be supported by an even finer field of some kind, a medium that didn't just permeate air, liquids, or solids, but one which saturated the entire universe itself. Tesla wasn't the discoverer of this field. In the nineteenth century it was already known about and accepted as fact by most of the scientific community. Not until the Michaelson/Morley Experiment of 1887, was it incorrectly dropped from scientific theory. The name for this all-pervading field was the "AETHER" (pronounced ether), and it can be described as a field of acoustic energy, or, to put it another way, it is made of 'sound'. Also, if Tesla was right, it appeared to be a field that could be engineered at will, as long as you had the correct key to unlock it.

Tesla's discovery revealed how the all-pervading Aether was full to capacity with *impulse currents*, and throughout the latter part of his career, these *impulse currents* became the foundation for his experiments

During this time, the inventor discovered that each pulse (individual pearl) gave the *impulse current* its frequency, and by changing the frequency, he could get different effects to take place on the same object. Furthermore, one could theoretically use this energy to manipulate solids, liquids, gases or even the vacuum of space, and get real physical effects to take place, regardless of distance.

Once Tesla understood the physical laws, he began to develop the equipment needed to harness and control the Aether.

Wireless Light Demonstration
In 1891, Tesla gave this demonstration before electrical engineers at
Columbia College. Even today, many believe Tesla was using high-
frequency alternating-currents to wirelessly light up his tubes. Yet
during that same period, he spoke publicly about the Aether and how
it could perform wireless transmission of power. The Serbian knew the
difference between the former, which could send wireless power short
distances, and the latter, which could revolutionize the world.

The modern Tesla Coil, as we know it today, is actually
based upon the work of the great English physicist Sir
Oliver Lodge, a man who spent much of his professional
life building machines trying to prove the existence of
the Aether. A quote from Gerry Vassilatos describes the
difference between Lodge's device, and Tesla's.

*"Tesla Transformers are not magnetoelectric
devices, they use radiant shockwaves, and produce pure voltage without current. No university High Frequency Coil must ever be called a
"Tesla Coil", since the devices usually employed
in demonstration halls are the direct result of
apparatus perfected by Sir Oliver Lodge and*

*not by Nikola Tesla. The Tesla Transformer is
an impulse apparatus, and cannot be as easily
constructed except by strict conformity with
parameters which Tesla enunciated."*

**–Gerry Vassilatos
Secrets of Cold War Technology-
Project Haarp and Beyond,
2000**

The modern Tesla Coil produces amazing spark discharges (lightning bolts) and electroluminescent effects that never fail to leave an audience in awe. But the Tesla-Magnifying Transformer, or T.M.T., was specifically designed for retarding the electric current, even though his machine could be tuned for producing side effects, like lightning. So, for the sake of simplicity, it is easier to call the modern Tesla Coil, Lodge Coil, and the original coils made by Nikola Tesla, the T.M.T.

A convenient way of visualising the difference between the two, is to view the Lodge Coil and T.M.T. like a sink full of water, with soap suds covering the top. In this analogy, the water is the voltage, while the suds are the electric currents. When the plug is pulled out of the sink, on a modern Lodge Coil, both the water and the suds go down the sink hole together, producing the amazing lightning shows we are all familiar with. On the other hand, the T.M.T. releases the water, but keeps the suds in the sink by plugging the hole only after the water is gone.

It is this basic difference that makes the T.M.T. so incredible. And to achieve that requirement, he used two specially designed coils ('primary' and 'secondary') to restrict the electric-currents within the wire, while letting the voltages flow over the top. To use another analogy, imagine Tesla's device acted like a musician's turning-fork, allowing him to change electrical energy into mechanical vibrations (resonance), and then use it to manipulate the environment.

Because the modern Lodge Coils release their electric-currents along with the voltage, the effects taking place only happen near the apparatus; with Tesla stating that "these effects were very weak". Conversely the inventor claimed that his *impulse currents* were extremely strong; indeed they had the power to be able to traverse great distances with ease.

In the ensuing years, Tesla made many public statements concerning his wireless endeavours, but for many of the scientists at the time, the Serbian was only dealing with High Frequency Alternating Currents, and to them this was nothing new. But that could not have been further from the truth, for the inventor of the Polyphase Alternating Current System knew the difference between his original electrical transmission system, which used wires, and this new system, which didn't. He even demonstrated his apparatus before these same people, but because of personal bias, they couldn't quite unlock the mental constraints that bound their minds.

CHAPTER FOUR

RESONANT POWER

"If you want to find the secrets of the universe, think in terms of energy, frequency and vibration."

–Nikola Tesla

THE ALL-PERVADING AETHER was the cornerstone of nineteenth-century science. It had the ability to explain many mysteries of the cosmos, yet there was no experiment delicate enough to directly detect its presence. It was a field that seemed invisible to the senses, yet it was Tesla, with his uncanny mental ability, who strove to unlock its unlimited possibilities.

If Tesla was alive at the beginning of the 21st century, he would be in total disagreement with what scientists believe concerning physics. The Serbian believed wholeheartedly in the Aether, and that atoms with their particles were nothing more than small dynamic fields within the all-pervading field. To use an analogy, imagine atoms and particles are little whirlpools, and the Aether is the ocean. Just as his previous quote suggests, the inventor believed that through the use of frequency and vibration, he could get the Aether to rise up and perform work at any point desired. This was the key to Tesla's Aether based inventions, including the greatest of them all, the Tesla-Magnifying-Transformer.

After more than half a decade of ardent experimentation into the Aether and how it could be manipulated, the inventor was ready to reveal to the public some of the results of those experiments. One of his more interesting public displays was performed before a handpicked audience of friends, friends of friends, and personally chosen news journalists. In this instance, the journalist was Chauncey Montgomery McGovern, and the article was for *Pearson's Magazine*.

The set of demonstrations performed by Nikola Tesla and his assistants were two-fold; first it gave a glimpse into what could be achieved by manipulating the Aether, and secondly it was to help draw in prospective investors, so the inventor could go to the next stage of sending radio signals and power around the world without wires.

THE NEW WIZARD OF THE WEST
By Chauncey Montgomery McGovern

"Not to stagger on being shown through the laboratory of Nikola Tesla requires the possession of an uncommonly sturdy mind. No person can escape a feeling of giddiness when permitted to pass into this miracle-factory and contemplate for a moment the amazing feats

which this young man can accomplish by the mere turning of a hand.

Fancy you seated in a large, well-lighted room with mountains of curious looking machinery on all sides. A tall, thin young man walks up to you, and by merely snapping his fingers creates instantaneously a ball of leaping red flame, and holds it calmly in his hands. As you gaze you are surprised to see it does not burn his fingers. He lets it fall upon his clothing, on his hair, into your lap, and finally, puts the ball of flame into a wooden box.

You are amazed to see that nowhere does the flame leave the slightest trace and you rub your eyes to make sure you are not asleep.

The odd flame having been extinguished as miraculously as it appeared, the tall, thin young man next signals to his assistants to close up all the windows. When this has been done the room is as dark as a cave. A moment later you hear the young man say in the laboured accentuation of the foreigner: "Now, my friends, I will make for you some daylight." Quick as a flash the whole laboratory is filled with a strange light as beautiful as that of the moon, but as strong as that of old Sol. As you glance up at the closed shutters on each window, you see that each of them is as tight as a vice, and that no rays are coming through them. Cast your eyes wherever you will you can see no trace of the source of the light.

Scarcely have you begun to marvel when the

light goes out by a touch on a button by the young man's hand. The room is in darkness again until the same laboured accentuation causes the reopening of all the shutters. Some animal is now brought out from a cage. It is tied to a platform. An electric current is applied to its body and in a second the animal is dead. The tall young man calls your attention to the fact that the indicator registers only one thousand volts, and the dead animal being removed, he jumps upon the platform himself, and his assistants apply the same current to the dismay of the spectators.

You feel a creeping sensation course up your back, and you see the indication slowly mounting up to nine hundred, and then one thousand volts and you involuntarily close your eyes, expecting the young man to fall dead before you the very next minute. But he does not budge. Quickly the indicator goes up, up, up, until presently it shows that ten thousand volts, then two million volts of electricity are pouring through the frame of the tall young man, who does not move a muscle. Nikola Tesla is holding in his hands balls of flames. At a sign, the current is stopped, the room is again made as dark as night, and presently the visitor sees the sharply defined black silhouette of the young man, with a beautiful halo of electricity in the background, formed by myriads of tongues of electric flame. The place is lighted once more, and as the young man comes up to you and shakes your hand, you twist it about in the same fashion as you have seen people do who hold the handles of a strong electric battery. The young man is literally a human electric 'live wire'. To tell of these and a thousand other wonders that Tesla does in a trice gives only a faint conception of their effect on the visitor. To really appreciate them one must see, hear and feel them in the flesh. It is a scientific treat of a lifetime, but it is a treat that few can enjoy, for the laboratory of Tesla is securely locked against everyone not provided with an introduction from a personal friend of the audacious wizard."

—Pearson's Magazine,
May, 1899

ELECTRIC OR AETHERIC

The public display written about in *Pearson's Magazine* is nothing less than startling. The author of the article was witness to an amazing feat of electrical engineering. The only downside was that while Tesla was only too happy to show off some of his latest innovations, the machinery that produced them would remain one of his most closely guarded secrets.

The question is, was this an ordinary electrical display, or was it some other form of energy that just looked like electricity? If Tesla's additional statements were correct,

Electric Fire Demonstration

Tesla states that it was just a parlour trick (in other words, it had no commercial value), but it must have been an impressive sight none the less. Known as 'cold-fire', the inventor was able to coat himself in electrical flame, then when he stretched out his arm, roaring tongues of fire would leap from his fingertips. If that wasn't strange enough, it is believed nearby objects would begin to bristle with rays, as well as emit strange sounds.

there could, in fact, be another form of energy creating the phenomena McGovern spoke of, and that it just happened to be the Aether. One of the experiments performed by Tesla for *Pearson's Magazine* involved the manufacture of artificial 'balls of flame' or what is sometimes called ball lightning. Even today, with all the information electrical engineers know, about electrical waves, they still don't have any idea how Tesla manufactured the phenomenon. Some of the present day eyewitness accounts concerning the natural manifestation of ball lightning, tell of its strange ability to be able to go straight through walls or windows of a house, and even the metal skin of a jetliner. The latter account was witnessed by the passengers and crew while flying at cruising altitude and caused no harmful effects on anybody while traversing long ways down the main cabin. The penetrative effects displayed by ball lightning are exactly how Tesla characterises the nature of *impulse currents*.

Tesla does however give us some clues as to the power source for ball lightning. He states, along with more recent researchers, how they could not just be absorbing their total energy from normal electrostatic origins, but there has to be some other agent imparting its power so it can appear as a glowing object. According to Tesla, the energy responsible for ball lightning is the Aether, and that's the reason it doesn't experience losses like normal electromagnetic waves do.

When properly harnessed, *impulse currents* allowed Tesla to do things that would normally seem impossible, including the three demonstrations performed for *Pearson's Magazine*.

VIBRATIONS

Another demonstration Tesla displayed for *Pearson's Magazine*, concerns the illumination of the lab with a powerful form of light. "This" the author said, "was as strong as the sun yet as beautiful as the moon". Glancing around the room Chauncey McGovern failed to detect a light source of any kind, and that included any rays which might be coming in from the sun outside.

When it comes to how the inventor achieved this feat, he states that he had found a way to get the atmospheric atoms in his lab to vibrate at unimaginable rates, forcing each one to produce its own excess light.

To quote from an interview with *The World* newspaper, Tesla provides a clear understanding of the importance of vibration and what it can do.

> *"The light of the sun, according to Mr Tesla, is the result of vibrations in 94,000,000 miles of ether* [Aether] *which separate us from the centre of this solar system. Mr Tesla is to produce here on earth vibrations similar to those which cause the sunlight, and thus to give us a light as good as that of the sun, with no danger from clouds or other obstructions. Mr Tesla has already achieved decided success in this line. He takes in his hand a long bar of glass, which, by vibration alone, lights up into most amazing brilliancy. He himself comes out of his experiments a most radiant creature, with light flaming at every pore of his skin, from the tips of his fingers and from the end of every hair on his head.*

In explaining his experiments, Mr Tesla uses figures calculated to pulverise an ordinary mind.

It is difficult for me, he said, to give you an idea that you will readily grasp about this question of vibration. In ordinary life our minds do not deal with the figures that come up in such investigations, but take a 5 and put after it 14 zeros, then you will have the number of vibrations which occur in the ether [Aether] every second and which produce light.

I carried out Mr Tesla's suggestion, with the following result—500,000,000,000,000.

All I have to do, said Mr Tesla, to duplicate the sunlight is to get this number of vibrations to the second with my machinery on earth; I have succeeded up to a certain point, but am still at work on the task."

–The World,
22 July, 1894

Tesla believed vibration was one of the agents responsible for changing the state of energy throughout the universe. His interview in *The World*, explained how he could get a dark room to light up like daylight by getting the atoms in the air to vibrate at 500 trillion times per second. But what was even more interesting is, he was only copying what happened naturally throughout space.

Tesla believed light was nothing more than vibrations in the Aether, and if light is the foundation for all physical reality, a simple formula of vibration rate would account for everything that we observe in the universe.

There were five years separating *The World* article from the *Pearson's* interview, so by the time the latter took place the inventor had perfected his illuminating device. Throughout his career, the New York inventor spoke about various applications his discoveries could be used for. These included manipulation of the atmosphere above the Arctic Circle so that a soft auroral type light could illuminate their long winter nights, or lighting a strip of atmosphere along the busy shipping lanes, giving maritime crews a well-lit route by which to sail at night.

RESONANT FREQUENCY

Soon after Tesla moved into his 46 Houston Street laboratory in 1895, nearby residents began experiencing unusual phenomena. Their curiosity became aroused when strange thumping noises, and light shows started emanating from his lab at all hours of the day and night.

Then by 1896, stories began to emerge about an unusual device the Serbian had developed called the Mechanical Oscillator. The Mechanical Oscillator was first developed because Tesla needed a high-frequency AC generator for a previous experiment he had conducted. It had an air driven rod that moved back and forth at incredibly high speeds, allowing the device to vibrate with such accuracy that it could move the mechanism in a clock with absolute precision.

Tesla also believed his Oscillator had the potential for therapeutic use. He would attach the device to a platform and study different effects upon himself as well as other people. On one occasion, while Tesla's good friend Samuel Clemens (Mark Twain) was visiting, he jumped at the chance to use it. Standing on the platform, the famous writer allowed the therapeutic vibrations to blissfully permeate through his body, thus taking no notice of Tesla's warnings about what long-term exposure might do to him. Continuing on, the vibrations coursed up through his body, causing his muscles, nerves and organs to release tension, and fall into a deep state of relaxation. The inventor protested again for the writer to get off the device or he would pay the price for his insubordination, but to no avail, Twain had nothing but praise for how this device could rehabilitate a worn out and tired population. Finally, as the inventor urgently stressed again for his subject to get off, the writer widened his eyes, while making his body taut, then snapped at Tesla "Where is it?", with Tesla pointing at the toilet and saying "Right over there through the little door in the corner." What Mr Clemens didn't understand was that the vibrations caused everything in his body to relax, including his bowel.

Continuing his experiments into vibrational effects upon different materials, Tesla decided to screw the Oscillator to one of the steel pillars in his laboratory. He then proceeded to start it up, causing the rod inside to oscillate back and forth at high speed.

By setting the device to the correct frequency of the steel beam, he brought both into harmonic vibration (resonance). Then as he increased the resonance, objects in his lab began to vibrate; increasing further, those initial objects stopped and others started, just like it was moving from layer to layer. Tesla knew what he was observing;

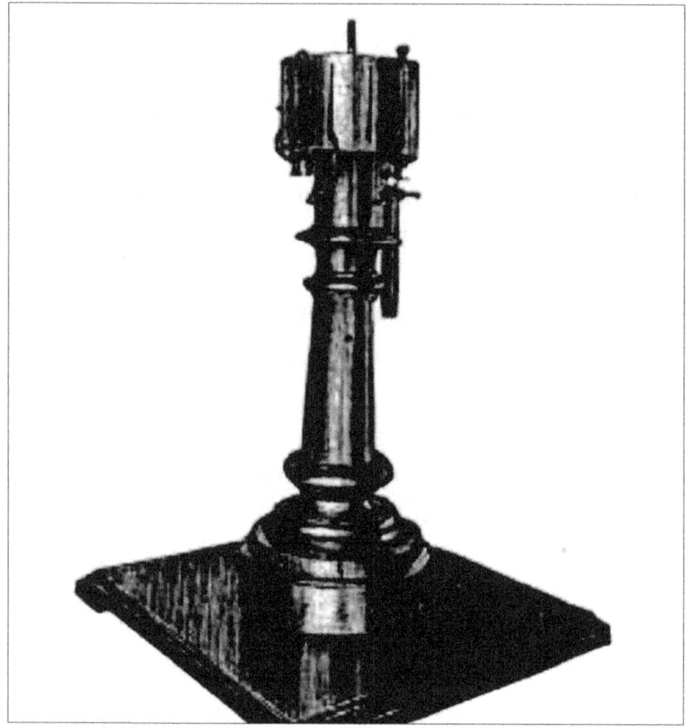

Tesla's Mechanical Oscillator

Developed and patented by Nikola Tesla, the Mechanical Oscillator was designed to vibrate at super high speeds, and with absolute precision. When attached, it is believed the device would combine with an objects resonant frequency and then manipulate it in various ways. Later in life, the inventor states how he had bolted one of his Oscillators to a steel beam in his lab, causing a build-up of vibrational resonance in the surrounding streets.

---

it appears each object had its own resonant pattern and when the vibrations reached a particular level it acted like channels on a radio, allowing him to individualize the frequency of the resonance.

Now, when it came to that area of New York city where Tesla had his laboratory, it was dominated by a range of buildings that included multi-storied tenement blocks, factories and even civic buildings, including a nearby police station. What Tesla didn't realize was, the small vibrations he experienced in his lab were minute compared to what was happening in the surrounding neighbourhood. That's because the steel superstructure of the lab, as well as the ground underneath, were acting like a giant tuning-fork, amplifying the vibrations as they went outward.

When it came to his experiments, Tesla usually liked to keep the locals guessing, but on that particular day, there was no mystery as to what the inventor was up to. It wasn't long before one of the officers from the nearby police station emotionally exclaimed, "That isn't an earthquake, it's that blankety-blank Tesla." As for the inventor himself, he too was coming to the realization of what was happening, especially when the shaking got so bad that heavy machinery started moving around the floor, while peculiar cracking sounds emanated around him. Now according to John J. O'Neill in his biography, *Prodigal Genius*, the police knew it was only a matter of time before their red brick building came crashing down around them. So, they stormed out of their headquarters, and down the street towards the inventor's workplace.

Throwing open the door, they see a tall, thin man swinging a sledgehammer at a small device on one of the pillars, thus smashing it and bringing the vibrations to an end. Laying the hammer down, the inventor denied any intentional guilt over the situation, even having the audacity to invite them back afterwards for a public demonstration.

Tesla called this field of science Tele-geodynamics, which loosely equates to resonance between two objects. The inventor does go on to make some other startling claims concerning the destructive potential of the Mechanical Oscillator. After being asked by a shrewd reporter from the *New York World – Telegram* about what it would take to destroy the Empire State Building, the inventor states:

> *"Five pounds of air."*
> *"Vibration will do anything. It would only be necessary to step up the vibrations of the machine to fit the natural vibration of the building and the building would come crashing down. That's why soldiers break step crossing a bridge."*

> *– Nikola Tesla*
> *New York World – Telegram*
> *1935*

Then there was his statement about how he could drop the Brooklyn Bridge into the East River in less than an hour. But that was surpassed by an even bolder statement in *World Today*.

> *"Tesla claims that in a few weeks he could set the earth's crust into such a state of vibration that it would rise and fall hundreds of feet and practically destroy civilization. A continuation of this process would, he says, eventually split the earth in two."*

> *– Nikola Tesla*
> *World Today,*
> *1912*

Tesla wasn't just making wild statements of fantasy; he fully understood the power inherent in resonance.

His description for splitting the planet states how the earth would heave hundreds of feet as the giant stationary waves moved around the planet. This mirrors an opera singer when she brings a fine wine glass into resonance with her voice. As she hits the right note, resonance is created and the glass begins to heave in and out, resulting in its destruction. The heaving effect can truly be appreciated when using a high-speed camera.

It is important to note that the device used to produce vibrations from five pounds of air would be nowhere near big enough to create earthquakes on a global or even a state-wide level, but regardless of its limitations, the physical principles of a global earthquake machine as stated by Tesla are still valid.

THE POWER OF SOUND

Sound is an extremely powerful agent for change, especially in the quantal world. If Tesla was correct and the Aether could be manipulated, then it would open up a well of possibilities for man.

We see many examples in the real world, (macro-world), where sound is used to manipulate physical objects; below are three examples of what can be done. The first example is where a small ball bearing is suspended (Levitation) in mid-air between two speakers that face each other. As each speaker makes its own wave, they couple together creating a stationary wave, then inside the stationary wave, a ball is suspended. In this experiment, it can be shown that through sound waves alone, mass can be suspended in place, as well as drawn in or propelled out.

In the second example, sound is used as a powerful tool for destroying physical objects. Known as lithotripsy, it is used in the medical industry for eliminating small kidney stones. In the procedure, a device projects sound waves (ultrasound) down through the skin and internal tissue without doing any harm. Then when it hits the kidney stone, it fractures it into small enough pieces so it/they can be passed through the urinary tract. This is an incredible example of the pinpoint accuracy, yet destructive power of sound.

When it comes to the manipulative power of sound, one of the most amazing inventions to ever appear was back in the nineties on a technology show called *Beyond 2000*. The host interviews a man who presents a small motor, like a Briggs and Stratton lawnmower engine. He proceeds to start it up, and as the motor comes to life, it sends out its annoying barrage of sound. It's at this point that the inventor flicks a switch on a little box on the side of the motor, which allows the engine to continue running but without any noise except for a slight whirr. The only clue he gave away was, when the sound wave came out of the motor, it was matched by a sound wave going in from the box, in effect cancelling out most of the noise.

That incredible little invention could have revolutionized our environment when it came to industrial noise pollution. Imagine that device being applied to a jet engine. The first thing to cross my mind was "when can I buy one?" But sadly twenty years later and it still hasn't been commercialized, so what became of it?

The three previous examples give us an important

glimpse into the power of sound, and even though the principles are related, normal everyday sound waves don't come close to the power Tesla was dealing with in the superfine Aether field. Tesla and many other scientists of the time, believed the Aether was responsible for all we observe in the universe, and it was only a matter of time before it could be harnessed for the benefit of man.

Tesla's experimentation into the physical properties of the Aether gave him a clear understanding of what was possible. But now it was time for him to prepare for the next stage of his plan, an undertaking that would lead him well away from the comfort and familiarity of New York city, and to the wild and remote town of Colorado Springs.

CHAPTER FIVE

COLORADO SPRINGS

"This planet, with all its appalling immensity, is to electric currents virtually no more than a small metal ball."

–Nikola Tesla

GUGLIELMO (PRONOUNCED GOOL-YELL-MO) Marconi was born on the 25th of April 1874, to an Italian father and an Irish mother. As a young man, he developed a keen interest in science and electricity, motivating him to take up studies at the University of Bologna. There he learnt about the theories of Heinrich Hertz (1857-1894), the German scientist who experimented with electromagnetic waves, and was honoured by co-sharing his name with EM waves, which are also called Hertzian waves.

Marconi's interest in EM waves led him into the world of radio and ultimately into sending wireless signals long distances. History may suggest that Marconi was the inventor of wireless telegraphy (transmission of messages without wires), but this account is more legend than fact. Marconi was more of a collator of other people's inventions, showing an uncanny ability for acquiring and improving the works of others, while finding the commercial potential for the finished product. It has to be said though, Marconi's sanitized version of plagiarism would end up making the Italian very wealthy, and one of the most recognisable names of the twentieth century.

The radio broadcast system has been one of the most beneficial devices ever invented. But what many historians are not familiar with however, is how Marconi obtained the components for his radio, and the irony is, when he revealed his invention to the world, he would stop Tesla's greatest invention from ever being commercialized.

Now 1899 proved to be a pivotal year for the electrical genius Nikola Tesla, specifically his development of wireless radio and power transmission to any point on the planet. After an exhaustive scientific investigation, the Serbian was finally ready to head out west and see if all he had laboured for, was worth it. Working out his destination carefully, he knew it had to be a location where the atmospheric conditions were favourable, and settled on a point where the Great Plains meet the foot of the Rocky Mountains.

Finally, in the month of May, with 30,000 dollars in his hand from friend and financier, John Jacob Astor (Waldorf Astoria fame), Tesla made his way to the growing town of Colorado Springs. The money would allow the inventor to perfect his T.M.T. on an industrial scale, which would enable him to bring it into resonance with the whole earth. So confident was the Serbian of his experimental results, that he knew it was only a matter of time before a commercial version would be unveiled to the world.

The main commercial reason he got on a train, and travelled half way across the continent to Colorado Springs, was because he needed secrecy, which was critical during this phase of his endeavour. He needed to stay ahead of his competitors who just might beat him to the punch, but in reality Tesla was so far ahead of the competition, including leading scientists, that they had little idea of what he was up too. Even though Tesla made carefully worded public statements concerning his work at the 'Springs', he was still extremely paranoid about information escaping his control; this even extended to those working for him, like his close associates Czito and Lowenstein.

More importantly, there are three main technical reasons why Tesla used Colorado Springs for the full-scale development of his equipment. The first was so he could get away from civilization, because his technology was still in an unproven state and could be extremely dangerous. That's why the inventor had to engineer many safety aspects into his giant transmitter before taking it back to the highly populated East Coast.

The second reason slots into the first; Tesla needed to get away from urbanization back east, but still required access to a constant supply of power to make his experiments viable. This is where Leonard E. Curtis comes in; it is his supply of electricity from the Colorado Springs Electric Company, that provided the inventor with all the power he would ever need. But just like many of Tesla's previous business partnerships, this one would also be thoroughly tested. As the story goes, during one particular experiment Tesla coupled his transmitter up with the Aether flow in the earth, but he underestimated the power locked up in his giant T.M.T. That energy then downloaded a voltage spike back into the town's main power supply, blowing out the generators at the power station. It is understood a very irate Curtis cut the Serbian's power until the inventor had fixed the generators.

The third reason Tesla needed to move to Colorado Springs was because of the amount of lightning that part of North America received on an annual basis. The Serbian had to find a place that had a frequent supply, so he could study how electricity flowed in and around the earth, but also why the two layers (earth and ionosphere) of the Schumann cavity interacted the way they did.

Making sure his laboratory was constructed to his exact specifications, it took Tesla about three months to complete the lab and its technical equipment, like the giant T.M.T. If it wasn't for the wooden tower, with its two-hundred-foot mast, and three-foot diameter copper ball on top, it could quite easily have been mistaken for a hay barn. But as the records state, when it came time for Tesla's assistant, Kolman Czito, to throw the switch, the barn-like structure revealed its true potential, and that included lightning bolts one hundred and thirty-five feet long, with their associated thunderclaps, heard fifteen miles away.

Once full-scale experimentation commenced, some highly unusual phenomena began to occur in the surrounding countryside. Some of them included small electrical discharges flowing from the ground to the shoes of bolting horses, electrostatic energy leapt off metallic objects, and butterflies were caught up in electrostatic type whirlwinds, which they could not escape.

While some of the broader aspects of Tesla's work at Colorado Springs are fascinating, it is impossible to recount all of the technical information. That's because a lot of it was buried in Tesla's head. What we do have is his public statements, his patents and his discourse with the *Electrical World & Engineer*, on the 5th of March, 1904. Much can be gleaned from the communique, with the first excerpt stating what some of Tesla's main goals were while based in New York city, as well as Colorado Springs.

> *"Toward the close of 1898 a systematic research carried on for a number of years with the object of perfecting a method of transmission of electrical energy through the natural medium led me to recognize three important necessities: first, to develop a transmitter of great power: second, to perfect means for individualizing and isolating the energy transmitted: and third, to ascertain the laws of propagation of currents through the earth and the atmosphere."*
>
> **–Nikola Tesla**
> **Electrical World & Engineer**

While in Colorado Springs the inventor worked out all facets of engineering needed for 'exciting' the earth; these included taking the power supplied by Curtis and transforming it into mechanical vibration. Then he would create the frequency needed to couple the energy to the earth or atmosphere. And lastly, he needed to know how these resonant waves manifested themselves throughout the earth and atmosphere, and more importantly how to protect the T.M.T. when they returned.

It has to be remembered that some of the terms Tesla used over one hundred years ago are not conducive with how we use those similar terms today. For example, the phrase 'Natural Medium' is a term Tesla frequently used when talking about the all-pervading Aether. Also in some instances when the inventor talked about 'Electrical' energy, he does not mean normal electrical current, but vibrations in the Aether.

In the next set of paragraphs from the *Electrical World & Engineer*, Tesla makes a fundamental discovery when investigating the lightning storms that frequent that part of the United States.

> *"It was on the third of July—the date I shall never forget—when I obtained the first decisive experimental evidence of a truth of overwhelming importance for the advancement of humanity. A dense mass of strongly charged clouds gathered in the west and towards the evening a violent storm broke loose which, after spending much of its fury in the mountains, was driven away with great velocity over the plains... Heavy and long persistent arcs (lightning) formed almost in regular time intervals.*
>
> *The recording apparatus being properly adjusted, its indications became fainter and fainter with the increasing distance of the storm, until they ceased altogether. I was watching in eager expectation. Sure enough in a little while the indications began again, grew stronger and stronger and after passing through a maximum, gradually decreased and ceased*

once more. Many times in regularly recurring intervals, the same actions were repeated until the storm which, as evident from simple computations, was moving with nearly constant speed, had retreated to a distance of about three hundred kilometres. Nor did these strange actions stop then but continued to manifest themselves with undiminished force.

No doubts whatever remained: I was observing 'Stationary Waves'.

As the source of disturbances moved away, the receiving circuit (wave-indicator metre) came successfully upon their nodes [wave cross-over point] and loops [peaks and troughs]. Impossible as it seemed, this planet, despite its vast extent, behaved like a conductor of limited dimensions."

–Nikola Tesla
Electrical World & Engineer

With his custom-built meters, Tesla wasn't recording the lightning itself, but the stationary waves (also called standing waves) that came off them. The lightning acted like a plunger moving through the natural medium (Aether), with the by-product being the stationary waves. The inventor not only discovered the natural processes that created the waves, but was going to imitate the process through his T.M.T. His intention was to send giant stationary waves into the earth and atmosphere, which would allow anyone with a tuned aerial to tap into.

The previous excerpt, together with the next one, shows the similarities when describing a stationary wave, and an echo. Tesla explains this simple analogy.

> *"That communication without wires to any point of the globe is practical with such apparatus would need no demonstration, but through a discovery which I obtained absolute certainty. Popularly explained it is exactly this: When we raise the voice and hear an echo in reply, we know that the sound of the voice must have reached a distant wall, or boundary, and must have been reflected from the same. Exactly as the sound, so an electrical wave is reflected, and the same evidence which is afforded by an echo is afforded by an electrical phenomena known as a "Stationary Wave"— that is, a wave with fixed nodal and ventral regions. Instead of sending sound vibrations toward a distant wall, I have sent electrical vibrations toward the remote boundaries of the earth, and instead of the wall the earth has replied. In place of an echo, I have obtained a stationary electrical wave, a wave reflected from afar."*

> **—Nikola Tesla**
> ***Electrical World & Engineer***

The echo analogy used by Tesla provides us with a description of how the giant stationary waves move through the natural medium (Aether), and that they were able to traverse the planet with relative ease. From his observations, the inventor knew if he could send a wave out with the correct frequency it would return to the giant T.M.T. with the peaks and troughs fixed in place, thus creating his much sought after stationary wave. Tesla goes on to say in his communique how the T.M.T. is not a Hertzian (electromagnetic) wave type apparatus.

> *"This is essentially a circuit of very high self-induction and small resistance which in its arrangement, mode of excitation and action may be said to be the diametrical opposite of a transmitting circuit typical of telegraphy by Hertzian or electromagnetic transverse radiations (waves).*
>
> *It is difficult to form an adequate idea of the marvellous power of this unique appliance, by the aid of which the globe will be transformed. The electromagnetic radiations (waves) being reduced to an insignificant quantity, and proper conditions of resonance maintained, the circuit acts like an immense pendulum storing indefinitely the energy of the primary (coil) exciting impulses and impressions upon the earth and its conducting atmosphere."*
>
> **—Nikola Tesla**
> ***Electrical World & Engineer***

The first sentence states emphatically how the giant

T.M.T. is not like the normal radio transmitting systems in use today. His system retards the electric-currents in the coils, while creating a condition in its circuitry for the easy flow of the voltage. Once the voltage is purified (or it could be said, transformed), it is pulsed back and forth between the two air and ground coil-transmitters. This is the 'pendulum effect' Tesla spoke about. That action allowed the appliance to impart impressions into and above the earth, which would be impossible for normal electromagnetic waves.

The last portion of Tesla's article in the *Electrical World & Engineer* states the superior capabilities of his apparatus compared to the Hertzian type radio system.

"As to the transmission of power through space, that is a project which I considered absolutely certain of success long since. Years ago, I was in a position to transmit wireless power to any distance without limit other than that imposed by the physical dimensions of the globe.

In my system, it makes no difference what the distance is. The efficiency of the transmission can be as high as 96 or 97 percent, and there are practically no losses except such as are inevitable in the running of the machinery. When there is no receiver there is no energy consumption anywhere. When the receiver is put on, it draws power. That is the exact opposite of the Hertz-wave system. In that case if you have a plant of 1,000 horsepower it is radiating all the time whether the energy is received or not. But in my system no power is lost. When there are no receivers the plant consumes only a few horsepower necessary to maintain the vibration."

–Nikola Tesla
Electrical World & Engineer

While based in Colorado Springs, the inventor conducted an experiment which involved the lighting of two hundred incandescent lamps, placed 26 miles from his laboratory. With each lamp consuming 50 watts, a total of 10,000 watts (13 horsepower) was being drawn out of the resonant stationary wave in the ground. He also states how distance was no barrier, as long as the T.M.T. was big enough to produce the waves needed to girdle the earth.

It wouldn't have mattered if the lamps were in Australia, Africa or Siberia, they would have lit up without any problems.

Another amazing aspect of the apparatus was the way in which it stored its energy. Tesla's other great invention, alternating current (AC), could not store its power the same way, only the fuel or water that drives the generators could be stored. Conversely, his giant T.M.T. was able to store its vibrations in the earth, and providing it remained running, the energy would stay there indefinitely. Only when someone stuck a tuned aerial in the ground, did some of the vibrational energy dissipate away.

One way to understand this concept, is to imagine the earth as a large water tank. When someone opens the valve to let some of the water out, only then does it begin to empty. It doesn't stop there though. Tesla states more than once throughout his career, that there is another important attribute these waves possess. That is, they are able to increase in power over time. The reason this is possible, is because they can absorb energy from the surrounding Aether, something an EM Hertzian wave doesn't do.

The unparalleled design of Tesla's T.M.T. at Colorado Springs is nothing short of genius. The earth and its atmosphere were treated like a tuning-fork, and the medium which carried the tuned vibrations was the all-pervading Aether.

Tesla's endeavours at Colorado Springs lasted for about a year. After that, he, his crew, and more importantly the wealth of knowledge he had gained at the 'Springs', headed back to the bustling east coast. That is where the inventor would fulfil his dream of setting up a whole new broadcasting industry, and not just for the transmission and remote-reception of radio signals, but more importantly, electrical power to any point on the globe.

If the Serbian born American was correct, and he could bring his ideas from the experimental, and into the commercial, then the world as we know it would never be the same again.

CHAPTER SIX

WARDENCLYFFE

"Electric power is everywhere present in unlimited quantities and can drive the world's machinery without the need of coal, oil or any other of the common fuels."

–Nikola Tesla

AFTER ALMOST A year experimenting in Colorado, Nikola Tesla returned to New York city to begin the next stage of his plan. It called for a giant version of his T.M.T. like the one he built out near the Rocky Mountains. If successful, the transmitter would become Tesla's crowning achievement, and give him the platform by which to expand the network worldwide, even allowing him to find new uses for it in the future.

On the 1st of March 1901, after acquiring the handsome sum of $150,000 from his old rival J.P. Morgan, the inventor feverishly set about drawing up the designs for his giant transmitter, and the location he chose to build it was a place called Shoreham, Long Island.

Combing the countryside for suitable land, the Serbian purchased 200 acres from James Warden. The reason he bought so much land was so he could eventually build a new metropolis around the tower, thus making Wardenclyffe the centre for world wireless communications and radio. Given the honour of putting his name on the facility, Warden was also employed as a sub-contractor on the site, constructing various service buildings around the tower.

It may have been Tesla's ideas, and Warden's land, but it was the financial and industrial tycoon J.P. Morgan who made it all possible. After purchasing 51% of Tesla's patents, the industrialist now had full control over the future direction of the project, with some even suggesting that the agreement would eventually lead to the venture's ultimate demise. Despite Morgan's paltry investment and overarching control, Tesla was desperate to forge ahead, so he wasted no time in preparing the foundations of the tower. His initial intention was to build one giant 600-foot high tower, which according to him, was all that was needed to make the system work on a global scale. Unfortunately, because of his eagerness to get started, the inventor over spent on land and equipment, eventually coming to the realization that his finances were dwindling quicker than he had predicted. Re-evaluating the situation, Tesla decided to construct a smaller tower, in the hope of one day building several others at different locations around the world. This change in direction allowed him to start building straight away, instructing his friend and architect Stanford White, to scale down the tower to 187 feet and draw up the new designs without delay. When it comes to the unique design of Wardenclyffe, what is commonly overlooked is the importance of an underground well system that connected the tower to the water table. Tesla's own quote describes the importance of this feature, and what it was designed to do.

> *"It is necessary for the machine to get a grip of the earth; otherwise it cannot shake the earth. It has to get a grip... so that the whole of this globe can quiver."*
>
> **–Nikola Tesla**

In view of Tesla's statement, it is obvious that the large well system under the tower was a ground antenna. It had several steel pipes drilled into the ground from the central well, giving the antenna the appearance of an underground fan, and thus allowing the T.M.T. to resonate the earth like a giant tuning fork. It was also necessary for the well system and tower itself to be lined with wood; which allowed the apparatus to build up potential without grounding out. It was crucial features like these which significantly increased the cost of the tower, adding credence to Tesla's statement that the apparatus was very expensive to build, but once functional, it was inexpensive to operate.

The unique design relates directly back to the energies involved, with Tesla stating how superior they were to EM waves.

> *"The practical significance of my system resides in the fact that the effects transmitted diminish only in a simple ratio with the distance, whereas in all other systems it is reduced in propor-*

> *tion to the square. To illustrate, if the distance*
> *be increased 100 fold I get 1/100th of the effect,*
> *while under the same conditions others can*
> *obtain at the very least 1/10.000th of the effect.*
> *This feature alone bars all competition."*

–*Nikola Tesla*

The above statement makes it clear that Wardenclyffe and the experimental station at Colorado Springs did not employ electromagnetic waves, but instead used waves of the Aether. The modern Marconi radio system is designed to emit large amounts of EM power through the antenna structure. This is necessary, because its strength drops off quickly as it leaves the antenna. The difference with Wardenclyffe, is that it was designed to retard the electric-currents in its coils, while using the voltage to vibrate the Aether in the earth.

If the transmitter was large enough, the Tesla system would have been able to send information, and even power, to the opposite side of the earth, with very little loss, and that loss was only in the conversion process, not in the transmission.

Another amazing feature of Tesla's transmitter was the unlimited amount of channels the system could employ, with author John J. O'Neill in his book the *Prodigal Genius*, stating that Wardenclyffe had 12 distinct applications it was going to benefit the world with.

> *" 1. Interconnection of the existing exchanges*
> *of offices all over the world.*
> *2. The establishment of a secret and non-*

interferable government telegraph system.

3. *Interconnection of all the present telephone exchanges or offices all over the Globe.*

4. *Universal distribution of general news, by telegraph or telephone, in connection with the press.*

5. *Establishment of a 'world system' of intelligence transmission for exclusive private use.*

6. *Interconnection and operation of all stock tickers of the world.*

7. *Establishment of a 'world system' of music distribution etc.*

8. *Universal registration of time by cheap clocks indicating the time with astronomical precision and requiring no attention whatsoever.*

9. *Facsimile transmission of typed or handwritten characters, letters, checks etc.*

10. *Establishment of a universal marine service enabling navigators of all ships to steer perfectly without compass, to determine the exact location, hour and speed to prevent collisions and disasters etc.*

11. *Inauguration of a system of world printing on land or sea.*

12. *Reproduction anywhere in the world of photographic pictures all kind drawings and records."*

–John J. O'Neill
Prodigal Genius

According to John J. O'Neill, Wardenclyffe tower was based solely around the transmission of radio, and was not designed to send useable wireless power. O'Neill, like many investigators, believed that because Wardenclyffe had to be down-sized, and the looming introduction of Marconi's plagiarized radio, Tesla had to concentrate on his broadcasting system and leave the wireless power till later.

Alternatively, other investigators believe Tesla was indeed attempting to send small quantities of power, to enable radios, clocks, stock tickers and other small apparatus, to function directly from Wardenclyffe, and eventually other towers in unison. In regards to this difference in opinion, Tesla's intentions for Wardenclyffe weren't exactly clear, but a statement by O'Neill does say the inventor had plans to build a far bigger transmitting tower in the future, and he already had an agreement in place with the Canadian Niagara Power Company to supply electricity. Tesla was so confident his facility would be built at Niagara Falls, that he stated it should be up and running in time to send wireless-power to the Paris International Exposition across the Atlantic.

There was still a lot of work for Tesla to do before Wardenclyffe was operational, and despite the inventor's confidence in his own ability, and trust in Morgan's generosity, different events were coming together to put an end to his dream of sending messages through the Aether.

Of those events, the most damaging was on the 12th of December 1901, when the letter 'S' was sent wirelessly by Morse Code for the very first time. That little act opened the way for instant communications across the world, allowing Marconi to claim the crown of being the first to send a message two thousand miles across the Atlantic, from Cornwall in England, to Labrador in Canada. This event made Marconi an instant celebrity around the world. As for Tesla, he naively thought Marconi would honour any primacy the Serbian held over radio, even stating.

> *"Marconi is a good fellow, let him continue...*
> *He is using 17 of my patents."*
>
> **–Nikola Tesla**

The Serbian's misplaced assurance in the patent laws led him to believe he would be given due recognition for his discoveries, and in turn receive the royalty payments he so desperately needed. Marconi didn't care. His plan was to divert all recognition away from the Serbian, and onto other notables like Maxwell, Kelvin and Hertz. He even states, how he didn't need Tesla's patents to demonstrate a working radio, but in reality, Marconi could not have given public acknowledgement to Tesla, as it would have meant admitting infringement upon the Serbian's patents. Tesla even said years later that Marconi had indeed admitted to plagiarism.

The absurdity of Marconi's fraudulent behaviour even drew the ire of the US Patent Office, who made a statement for one of Marconi's patent applications in 1903.

> *"Many of the claims are not patentable over*
> *Tesla patents Nos 645,576 and 649,621, of record,*
> *the amendment to overcome said references as*
> *well as Marconi's pretended ignorance of the*

nature of a 'Tesla oscillator' being little short of absurd... the term 'Tesla oscillator' has become a household word on both continents (Europe and North America)."

–US Patent Office

The situation between the two escalated to the point where the courts became involved, driving Marconi and Tesla into a media-driven feud that would last for the rest of their lives. In fact, it would be six months after Tesla's death in 1943 that the US Supreme Court would rule in the Serbian's favour, overturning Marconi's primacy for the invention of radio. One wonders if it was sheer coincidence why the US Supreme Court waited so long to see justice done, or was it Morgan's overarching dominance of all things industrial that brought the legal establishment to heel. It is known that Morgan was present when the broadcasting conglomerate RCA came into being, which just happened to use Marconi's plagiarized radio system.

Even after Marconi's radio was revealed to the world, Tesla continued on with construction of Wardenclyffe. Unfortunately, finances were drying up fast, so in desperation, Tesla approached Morgan in an attempt to release more of the money he had promised the inventor. Had it not been for his reputation, Tesla would not have got as far as he did with Morgan, but even he was beginning to have trouble holding the tycoon's attention. Some say it was Morgan's control over radio, while others point towards Tesla's constant changes to Wardenclyffe as the reason for the industrialist's growing disinterest. Whatever it was,

Tesla had no choice but to continue on with the project. He had to prove to Morgan the tower was technologically feasible, but of course, without his money how was he going to achieve this.

It is at this point, so the story goes, that Tesla revealed his true intentions for Wardenclyffe, including future developments, like the full-scale wireless power transmitter at Niagara Falls. But is that true? One thing favouring this scenario is, it would have been okay to inform Morgan of his plans to transmit Aether radio waves, but it would have been disadvantageous for the inventor to reveal its full potential for sending wireless power. Once Tesla revealed his master plan, the Wall Street tycoon would have sealed the Wizard's fate, and according to the memos that is exactly what he did. Morgan's reach was vast. He had strangled the AC patents away from Westinghouse, almost bankrupting the company in the process, and at the same time building General Electric up into one of the corporate giants we know of today. It wasn't all plain sailing for Morgan though; the industrialist was having his own problems within the corporate world. Some of them included the 'Run on the Markets', which caused violent swings within the American economy, then there was the downward spiral in share prices for one of his other great industrial concerns, Northern Pacific Railroad. Those events caused Morgan to feel threatened by what was happening around him, and he didn't need an inventor with tunnel-vision pressuring him for additional finances, especially now that Tesla wasn't the only radio innovator around.

Despite those events, most researchers believe it was only after Marconi successfully transmitted his message across the

Atlantic, that Tesla started having trouble getting an audience with Morgan. Now, the seriousness of the situation grips Tesla, and it isn't long before the inventor begins pleading with the industrialist in a series of memos, hoping to save his facility from financial ruin. Tesla's desperation begins to show in one such memo to Morgan.

> *"Won't you enable me to complete the work and show you that you have not made a mistake in giving me a cheque book to draw on your honoured house? If you will imagine that I have found the stone of the philosophers, you will not be far from the truth. My invention will cause a revolution so great that almost all values and all human relations will be profoundly modified."*

> **–Nikola Tesla**

By this time, Morgan's attitude towards the inventor is starting to show, and in the next quote, it only gets worse.

> *"Had you simply achieved what I had asked, you would not be in this predicament."*

> **–J.P. Morgan**

Continuing his dialogue with the powerful industrialist, Tesla states that no one else can do what he can, telling Morgan that he controls all intellectual rights.

"Mr Morgan, I beg to call to your attention that my patents control absolutely all essential features and that my work is in such a shape that whenever you tell me to go ahead, I shall girdle the globe in three months as surely as my name is Tesla. I have promised to the St Louis people to open the door of the exposition with power transmitted from here. It is a great opportunity, Mr Morgan, I can easily do it, but if you do not aid me soon, it will be too late."

–Nikola Tesla

On the 13th of January 1904, after exchanging various memos, Morgan makes the situation clear.

"My dear sir
In reply to your note I regret to say that I should not be willing to advance any further amounts of money as I have already told you. Of course, I wish you every success in your undertaking."

–Yours truly, J. Pierpont Morgan

Morgan's attitude towards Tesla is now clear; the industrialist has no need of the inventor's intellectual ability. But it didn't stop there; a strange memo states how there was more to the situation than was previously thought. Tesla writes to Morgan on the 22nd of January, 1904.

"… are you going to leave me in a hole?!! I have made a thousand powerful enemies on your account, because I have told them that I value one of your shoestrings more than all of them... In a hundred years from now, this country would give much for the first honours of transmitting power without wires. It must be done by my methods and apparatus and I should be aided to do it first myself."

–Nikola Tesla

Tesla was very well known by this time, and there were plenty of wealthy people willing to see the inventor succeed with Wardenclyffe. Unfortunately for Tesla, apart from a couple of small investments, the larger investors shied away once they knew who the business partner was.

Morgan's influence and power allowed him to isolate Tesla from future investors, with a memo from the Serbian begging the capitalist to either help finish the complex at Wardenclyffe, or relinquish his 51% stake so the inventor could find other investors. History tells us, Morgan would never relinquish his controlling share, even if it meant the loss of his $150,000 investment.

Morgan knew all too well, the Serbian was more than capable of fulfilling his ambitions, and it was more than likely the capitalist went out of his way to stop Tesla ever receiving any alternative finance in the future; remember, Morgan was a part of the RCA cartel. While Morgan projected a fierce reputation around Wall Street, when he died people were quite surprised to discover his net worth was a lot smaller than had previously been thought; it has been suggested by some that the industrialist was nothing more than a front for old European money. So, for families like Morgan, Rockefeller, and Carnegie who were collectively cornering commodities such as power, oil, steel, transportation, and heavy industry, they couldn't allow a disruptive technology like Wardenclyffe to survive. That's because it could destroy all the robber barons had worked so hard to accumulate, and the irony is, it was Tesla's invention of AC power that gave these financiers the means by which to expand the way they did. Despite his fighting spirit, it was over for the inventor; his dream of girding the earth with giant stationary waves was never to be, Tesla had tried his best to extend Morgan's financial support, but the New York industrialist refused to support him in any way.

It got to the point where the ordeal wasn't just effecting Tesla's finances, it also impacted on his mental health. He had sunk to such a low that he wrote this memo to Morgan.

"Since a year, Mr Morgan, there has been hardly a night when my pillow is not bathed in tears, but you must not think me a weak man for that."

–Nikola Tesla

It wasn't just Tesla who was concerned about his own health; other people had equal concern for the inventor's mental wellbeing. Tesla's friend/manager, George Scherff, states in 1906:

"I have scarcely ever seen you out of sorts as last Sunday; and I was frightened."

–George Scherff

The Serbian's emotional recovery would come with time, helped by his ability to develop new ideas quickly, however, his dream of completing Wardenclyffe would never cease; he talked about the tower until the end of his life. The period between 1903 and 1917 made sure the inventor would never see his dream come true. Some of those events are mentioned below:

- The 'Rich Man's Panic' was an economic downturn that hit the country in 1903, causing many new prospective investors to dry up permanently.
- The Colorado Springs Power Company sued Tesla in 1904, for electricity used by the inventor while testing his theories in the remote Rocky Mountain town.
- By mid-1905 Tesla's fundamental patent rights expired on his alternating current motors and associated applications.
- Tesla had electrical equipment sold-off in 1906, because of unpaid bills.
- Following Morgan's withdrawal of support, Tesla tried to develop other inventions. His intention was to build up his finances so Wardenclyffe could be restarted, but his endeavours failed to materialize.
- By 1912, creditors had taken control of

Wardenclyffe, after the inventor had pretty much exhausted all other avenues.

His financial difficulties hadn't just wrecked his career, they had also affected his lifestyle. One account in 1915 describes how Tesla was living at the plush Waldorf-Astoria, appearing to live the good life. But as Court documents show, the inventor hadn't paid his rent for quite some time, causing him to amass a debt of around $19,000. Having lost patience with the famous Electrical Wizard, the proprietor of the Waldorf, George C. Boldt, put pressure on Tesla to relinquish control of Wardenclyffe. This was so they could sell equipment to recuperate some of the debts. The records do say how Tesla kept a small interest in the tower, hoping to some-day resurrect it, but sadly this was never to be.

Because the tower was so close to completion, a legend states, that in an act of defiance the inventor proceeded out to the great tower for one final display. Several nights of atmospheric bombardment threw the small area of New England into wonderment, giving off giant lightning bolts with their associated thunder claps; it is also believed the locals experienced the formation of Auroras above the site. Neither the inventor nor anyone else claimed responsibility for this show-stopping event, but one thing is for sure, the giant wireless broadcasting tower at Wardenclyffe would never display its secrets again.

In 1917, papers at the time said the US Army had demolished Wardenclyffe during the Great War, believing that German U-boat commanders were using it as a landmark for suspected incursions. But the truth is quite different: counter to Tesla's vehement protests, the tower was pulled down under the orders of Waldorf management. And because the tower was exposed to the elements for many years it was too dangerous to dismantle by hand, so it was unceremoniously blown up with dynamite.

The story of Nikola Tesla doesn't stop with Marconi's plagiarism, and Morgan's underhandedness, there appears to be an outright denial of his contribution to the field of science. Edison's complicit backing to omit Tesla's name from the electrical textbooks of the time, showed how there was a concerted effort to keep his theories out of the mainstream. These historical records have never been updated, even to the present day; consequently, Tesla has never received appropriate recognition in the field of electrical theory. The Serbian declares to Morgan concerning the way he is being portrayed.

"Mr Morgan, thanks for seeing me. My enemies have been so successful in representing me as a poet and visionary [dreamer] *that it is absolutely imperative for me to put out something commercial without delay."*

–Nikola Tesla

Tesla's concern at the way he had been portrayed publicly had merit: all the evidence points to the fact that the threat of Tesla's wireless power could have overthrown all existing electrical power (as well as gasoline) systems, which meant Morgan and his kind had little option but to protect their own interests now and in the future, even to the extent of rewriting the history books.

LATER YEARS

In the years following the demolition of Wardenclyffe, Tesla went on to develop many other inventions; some of them included applications for his bladeless turbines, designing VTOL aircraft, and provocative statements about a mysterious machine he called his Death Ray. There were also less well-known inventions he built while working for companies such as Budd National and Allis Chalmers. On one occasion, while working for the Waltham company, the inventor developed the Air-Friction Speedometer for vehicles. This reinforces the notion that Tesla's fertile mind was still developing new ideas late into his life, refuting the commonly held misconception of how he was a broken-down recluse.

It is also well known that during this period Tesla had developed business relationships with some of the great industrialists of the time. One of these was John J. Hammond, a defence contractor with the US Military. What is less well known though, is that the inventor had likewise developed mysterious associations with foreign European governments. Make no mistake—Tesla loved his adopted country, the United States, he also loved its "can do" attitude, which allowed free thinkers like himself to take their ideas and commercialize them for the benefit of all. But despite this, it is known he had tried to establish commercial links with people who were already in, or coming under the heavy influence of communism and fascism, but it has to be remembered that Stalin's Russia was still considered an ally by the West, and likewise Hitler hadn't started World War II.

One reason why Tesla approached foreign governments for help, was because he was being treated with contempt by western governments, in particular, the United States. They just ignored him when he requested to see them regarding any new breakthroughs he had developed. Some say the US Government became disinterested because they thought the inventor had become a crackpot, but this argument is countered by the fact that military-industrialist John J. Hammond desired to know the technical details of Wardenclyffe, which Tesla denied, declaring how Morgan still held majority ownership over his patents.

In his later life, Tesla fought many patent infringements around his intellectual property rights, including law suits against some of the biggest industrial corporations on earth—even his old ally the Westinghouse Corporation, which long after George Westinghouse's death in 1914, had the audacity to plagiarize the inventor's radio patents. In this murky period of the company's history, some say a settlement was reached between Tesla and the Westinghouse Corporation, which allowed the inventor to live out his days at the Hotel New Yorker, rent-free.

It wasn't just Tesla's career that was multifaceted, it was also his complex personality and set of phobias which made him such an intriguing character. Those around Tesla, knew he was an individual who prided himself on discipline and self-control, and nothing epitomized the man more than his work ethics. It was known the inventor would work multiple shifts back to back until he was so exhausted that he would go to sleep in the lab for short spells, and then like clockwork start all over again.

That discipline also extended into his personal life, where he denied himself marital love and affection, for the purpose of dedicating himself to his inventiveness. All this despite being known as a ladies-man, with many well to do women of the day trying to court him, including J.P. Morgan's daughter.

He may have demonstrated great self-control, but that didn't help the phobias he displayed in his character. An example can be seen with the unrealistic penchant he had for cleanliness, demanding a particular table be reserved for him alone while staying at the Hotel New Yorker, or only using a towel once, not shaking hands, and having the table reset if a fly inadvertently landed on it. Another bizarre phobia was his extreme loathing of women's jewellery, especially pearls.

Some suggest the symptoms for his obsessive-compulsive behaviour started after his long-running dispute with Morgan left him emotionally fragile for a time. In my opinion, his personal issues took nothing away from his inventive prowess, and to the contrary, it just adds to the legend that is Nikola Tesla.

By the end of 1942, Tesla's health was beginning to deteriorate, his age and strict eating habits made him look skeletal in appearance, and feeble in movement. The inventor was now a shadow of his former self, and it seemed the final curtain was coming down on one of the greatest geniuses the world has ever known.

On the 7th of January 1943, Nikola Tesla died at the age of 86. The story goes, that after two days of showing the 'Do Not Disturb' sign on his door, the maid Alice Monaghan, went in and found the Serbian's body on the bed; he had died in his sleep.

Such was Tesla's renown, that on the 10th of January Fiorello La Guardia read his eulogy over the radio, it befitted the man who had invented the device the mayor was speaking into. Then on the 12th of January, Tesla received a state funeral at the Cathedral of St. John the Divine in New York city. In attendance was a large crowd of 2000 people, including many dignitaries from the financial and political worlds. At the conclusion of the funeral service, Tesla's body was taken to Ferncliffe Cemetery, where the great inventor was cremated.

Not until 1957, were Tesla's ashes placed into a metal sphere, (his favourite shape), and sent to Yugoslavia to be put on display at the Nikola Tesla Museum in Belgrade.

THE CONTROVERSY

The story of Nikola Tesla doesn't end with his passing, but there was a lot of controversy surrounding the period just before and after his death. Some have claimed the Serbian was whisked off to some far-flung exotic location where he developed his ideas to the fullest, while others have suggested he was murdered because he was about to go public with some of his most profound secrets. Putting aside some of the more fanciful theories concerning Tesla's final movements, I think it is the scenario laid out by author Margaret Cheney in her book *Tesla, A Man Out of Time*, that provides the best evidence for events around the inventor's death.

Cheney describes how a series of memos between three US government departments led to a trail of interest, that went all the way up to the highest echelons of US Military Intelligence.

Tesla's collection of papers, which weighed in the tons,

were placed in boxes and barrels and stored at different locations around New York city. According to the records, the Federal Bureau of Investigation (FBI) ordered Tesla's estate to be handed over to the Office of Alien Property (OAP). Their reasoning was because the inventor didn't have a will, and the contents could have ended up anywhere. This is where Tesla's nephew Sava Kosanovic comes in. He was the Yugoslavian Ambassador to the US at the time. Through the courts, he ordered the contents of Tesla's estate to be taken to a storage facility in Manhattan, and from there to the inventor's native country.

Where the whole situation gets murky is around a three-way set of memos sent between the Office of Alien Property, Air Technical Service Command at Wright Field (Wright Patterson Airforce Base) in Dayton Ohio and Military Intelligence in Washington. If correct, this proves that the US Government was indeed interested in Tesla's papers, and made it clear, they wanted to see them before anyone else, especially the Yugoslavian Government. On the 22nd of January 1946, a memo is sent to the OAP from Col. Ralph E. Doty at Military Intelligence:

> *"This office is in receipt of a communication from Headquarters Air Technical Service Command, Wright Field, requesting that we ascertain the whereabouts of the files of the late scientist Nichola Tesla, which may contain data of great value to the above Headquarters.*
>
> *It has been indicated that your office might have these files in custody. If this is true, we would like to request your consent for a representative of the Air Technical Service Command to review*

them. In view of the extreme importance of these files to the above command, we would like to request that we be advised of any attempt by any other agency to obtain them."

–Col. Ralph E. Doty

That memo was sent by a liaison officer because of its perceived importance. Then in 1947, another interesting memo is sent by a David L. Bazelon from the OAP, to Wright Field in Ohio:

"Our records do not reveal that this material has been returned."

–David L. Bazelon

In response to this request a Colonel Duffy, who was Chief of the Electronics Plans Section at Wright Field, states:

"These reports are now in the possession of the Electronics Subdivision and are being evaluated... at that time (Jan 1 1948) your office will be contacted with respect to final disposition of these papers."

–Col. Duffy

According to Margaret Cheney, the papers were never returned to the OAP, and were missing when Tesla's estate was sent to Yugoslavia. Also, Marc Siefer, in his book the *Wizard*, states how the FBI, through its Chief J. Edgar Hoover, kept a close eye on the Tesla papers because of Kosanovic's ties to a foreign government, one that became communist in 1946. Siefer reveals that the OAP was the only US agency with the jurisdiction to legally hold the papers so they wouldn't be released to Tesla's nephew. That gave other US agencies time to inspect them for sensitive information that might help the US government in a time of war, as well as into the future.

While mystery still surrounds the movement of Tesla's papers, one thing is certain, various groups were extremely interested in them, and secrecy was of the utmost importance once they were obtained.

The story of where Tesla's papers went, and what they were ultimately used for, requires a more in-depth analysis than what is written in this chapter, and because of its incredible potential as a military weapon, any evidence proving its existence must also be mentioned. In *East Meets West* (chapter 11), a case is presented concerning the possibility of who might be in possession of the technology, and why their true identity has been so hard to uncover.

When it comes to Tesla's more esoteric work, it is difficult to separate fact from fiction, and that is not helped by the lack of understanding surrounding his invention, the T.M.T. While many know of its use in wireless power, few fully understand the potential his technology possesses. There were even newspaper reports at the time, revealing how Tesla was working on many other Aether based inventions, and if they could be commercialised, they would benefit our lives immeasurably.

In the next chapter, some of the technologies and theories Tesla spoke about will be presented, which may, in fact, help us understand why his revolutionary technologies have never seen the light of day.

CHAPTER SEVEN

VISIONS OF UTOPIA

"Let the future tell the truth and evaluate each one according to his works and accomplishments. The present is theirs; the future, for which I really worked, is mine."

–Nikola Tesla

THERE IS NO denying the world has seen a profound technological transformation over the last two hundred odd years. The advancements in information, transportation and food production amongst many others, attest to the speed at which the world has advanced in every area of our lives. We have seen a man walk on the moon, and space probes travel out to the edge of our solar system. Conversely, we've looked deep into the human body and even the atom itself, and all of this with the help of the electromagnetic wave.

While the industrial revolution has profoundly affected mankind in both negative and positive ways, what if there had been another discovery that used Aether waves instead. If men like Nikola Tesla, T. Henry Moray, T. Townsend Brown and Royal Raymond Rife had succeeded in their endeavours, the dream of an Aetheric revolution could have ushered in an unprecedented time of peace and prosperity.

The article below shows just where mankind was heading during that period of unrivalled innovation and free thinking. Among those innovations was the promise of wireless power, an energy system which had the potential to improve the lives of people in a profound manner.

MARVELS OF THE TWENTIETH CENTURY

"Those of us whose privilege it was to witness the glorious sunset of the old and the radiant dawn of the new century must have been impressed by two things—the marvellous accomplishments of the dying era and the mysterious possibilities locked in the heart of the new born child of time. The Nineteenth Century produced many wonderful demonstrations of man's supremacy over his surroundings. He had annihilated distance, captured sound, pictured motion, and chained the lightning. Things which a hundred years before would have been regarded as weird and unnatural had come to be looked upon as commonplace. Great inventions, though they might excite curiosity, interest and admiration, had ceased to startle. Yet human credulity must have its limit, and it would seem as though Nikola Tesla, the Electrical Wizard of the West,

had dared to step perilously close to the border. He has tamed and unchained the lightning. He has given us wireless telegraphy, and he claims that in time, not very far distant, messages will be sent across the ocean, without cables and at much lower cost than at present. His great accomplishments, however, pale into insignificance before his predictions of wonderful inventions yet to come. According to a paper in the 'Humanitarian', Tesla regards the transmission of electricity without wires as the all-surpassing task of the engineer, the practical consummation of which would mean that energy would be available for the uses of man at any point of the globe, in quantities virtually unlimited.

"Export of power would have become a chief source of income for many happily situated countries, as the United States, Canada, Central and South America, Switzerland and Sweden. Men could settle down everywhere fertilize and irrigate the soil with little effort, and convert barren deserts into gardens, and thus the entire globe could be transformed and made a fitter abode for mankind."

**–Reported in the Wanganui Chronicle,
8th October, 1901**

No area of our lives was left unaffected by these revolutionary ideas and inventions, everything from radio to lighting, and transport to agriculture, was advancing at an unprecedented rate. But what about Aether technology, could it have given us a whole new industrial infrastructure? From what Tesla had stated, it appears there were many esoteric discoveries waiting to be revealed. The following quotes in this chapter only mention a portion of what Tesla spoke about, and up until the end of his life, he believed wholeheartedly that the Aether was the driving force behind those discoveries.

RADIATION

"In my papers published in the Electrical Review in 1896 and 1897, several years before the discovery of radioferous ores, I described radiations of this sort... and I still hold to the opinion, believing Radium to be due to a process akin to combustion in which not oxygen but an all-pervading medium much finer involved..."

–Nikola Tesla
Reported in the Oamaru Mail,
15th June, 1912

Through his own experimental observations, Tesla did not believe radiation was a product of atomic decay. He believed atoms like Radium produced excessive radiation by converting large quantities of incoming Aether into visible EM energy. The inventor had previously stated that if the atom could be shielded from the incoming Aether currents, the radiation process would cease. If that is correct, then it implies EM-radiations manifesting out of atoms are the result of Aether-conversion, not atomic-decay. It's interesting to note how the first man to officially split the atom, New Zealander Lord Ernest Rutherford, was a keen admirer of Nikola Tesla and his theories.

If Tesla was correct, then radioactive detritus from places like Chernobyl could quite easily be purified, but the irony is, if wireless power was available to the public, we wouldn't need nuclear power plants in the first place.

PRIMARY SOLAR RAYS

"According to my investigations, the sun emits a radiation of such a penetrative power that it is virtually impossible to absorb it in Lead or other substances... This ray which I call the 'primary solar ray', gives rise to a secondary which now is commonly called the 'cosmic ray'."

–Nikola Tesla
Reported in the Dunstan Times,
21st September, 1931

Tesla believed that all the stars propel vast quantities of energy he called 'primary solar rays' into space, but they are not the same cosmic rays mainstream science talks about. He states how the sun was the greatest local source of this energy, but it still flowed down to the earth's surface at night, suggesting that it was different to regular EM energy. From his statements we know he wasn't referring to modern day solar cell technology, but through the use of his patent No 685,957, he maintains he could harvest this inexhaustible supply of energy for man's benefit.

SUPERLUMINAL

"More-over, the messages being sent underground, any possibility of interference is obviated. Mr Tesla's claims are very interesting, and that is why they are given such prominence. He boldly asserts that distance is no obstacle... To an interviewer Mr Tesla said that he had proved from a station he had already established that the very powerful current developed by the transmitter traversed the entire globe, and returned to its starting point in an interval of eighty-four one thousandths of a second, this journey of 25,000 miles being effected almost without any loss of energy..."

–Nikola Tesla
Reported in the New Zealand Herald,
8th January, 1910

The inventor declared how *impulse currents* were not bound by the speed of light, but moved superluminally. He also said how some of the waves he sent around the earth, had average speeds of one and a half times the

300,000 km/sec speed of light, which is 450,000 km/sec. That wasn't all, at certain points on the globe, their velocity sped up to infinity.

As mentioned before, this would allow for massive advances in communications, computer speeds, artificial intelligence and a myriad of other applications which would benefit the world greatly.

POWER MAGNIFICATION

"The design which I have adopted, will have a transmitter which will emit a wave complex of a total maximum activity of 10,000,000 horse-power, one percent of which is enough to girdle the globe. This enormous rate of energy delivery—it is twice as much as the force of Niagara Falls—is obtainable only by the use of certain artifices which I shall make known sometime in the future.

We have been offered ten thousand horse-power from the Canadian Power Company. What I want to do is to build machinery there and transmit this power to different parts of the globe..."

–Nikola Tesla
Reported in the Star,
4th March, 1905

Statements like this brought the inventor much ridicule from scientists as well as the press, since he intended to transmit more energy than what Niagara Falls could provide. What they didn't realise, was the inventor had discovered a very unusual property associated with *Aetheric-impulse currents*, a property Hertzian waves didn't possess. In simple terms, it is believed that Tesla was able to send a wave-complex into the ground and pull extra power in from the surrounding Aether field. One way to visualize the process is to imagine a snowball rolling down a snow-covered hill. As it moves it captures more snow, and thus grows in size. Known as a Power Tap, this is not perpetual motion, but instead, the wave-complex draws energy in from the all-pervading field that is already there.

One obvious benefit is excess abundant energy for the consumer. But that's not all, the process could also be used on other rocky planets, giving rise to the possibility of terraforming on a mass scale.

IMPULSE WAVE TUNING

"He can so regulate the intensity of the vibrations that he can transmit through the tissues of his body a force of something like 40 or 50 thousand horse-power. He gave an illustration of this sort to some learned society—to their horror—when the current that passed through his body melted the wire that joined his hands. Yet he suffered no inconvenience."

–Nikola Tesla
Reported in the Southland Times,
9th August, 1900

Another interesting discovery made by the inventor, was when he held one hand on a coil while holding a bar of steel in the other. When the coil was switched on, the waves would pass harmlessly through his body and into the steel, causing it to melt or even explode from the end.

This is not normal melting by heat, otherwise, his hand would have been consumed in a few seconds. Instead, it used resonance to induce molecular change, thus causing the metal to become syrup like in its condition. Called Cold Moulding, if this technique was used in the manufacturing industry, materials like Tungsten and Titanium could quite easily be reformed into any shape without the need for expensive smelting furnaces.

KINETIC RESONANCE

In author John J. O'Neill's book, *Prodigal Genius*, he explains how Tesla could get objects to vibrate, move, and even explode remotely through resonance.

By simple 'remote-tuning', certain objects in the room would perform while all others were at perfect rest. Besides its military potential, one civilian application would be its use in finding and assessing objects hidden from sight.

FORCES ARE ENGINEERABLE

*"I have worked out a dynamic theory of gravity
in all details and hope to give this to the world*

very soon. It explains the causes of this force and the motions of heavenly bodies under its influence so satisfactorily that it will put an end to idle speculations and false conceptions, as that of 'curved space'."

–Nikola Tesla,
1939

Totally opposed to Einstein's Theory of Relativity, Tesla states how he had conducted experiments into manipulating the force of gravity upon objects. He mentions more than once, he had formulated a complete theory of gravity by the 1890s, and had called it the Dynamic Theory of Gravity. Once all the necessary patents were obtained, it was the inventor's intention to release his theory to the world, but to this day it has never come to light.

If Tesla had indeed achieved this feat, the ramifications for mankind would be nothing less than the full-scale exploration of the Universe through what is commonly referred to as Anti-Gravity.

FORCE-FIELDS

"Some scientists have maintained that they are merely optical illusions, and so Tesla himself thought until they began to appear accidentally on his high-voltage equipment... In modern plasma physics the most commonly held theory is that the fireball receives its energy

from its surroundings by a naturally created electromagnetic field."

–Margaret Cheney
Tesla. Man Out Of Time

"First, Tesla claimed that the 'lightning balls' (which destroyed his equipment) could be used to destroy aircraft."

–Mr Clarence Kelley, FBI Director,
3rd July, 1980

Tesla may never have observed natural ball lightning, but he did say he could manufacture them. While the phenomenon is still largely misunderstood by modern physics, both it and Tesla do agree ball lightning receives its energy from an outside source. If this indeed is the case, then his Aetheric experiments at Colorado Springs most likely holds the key to their power source.

Tesla does give us some clues as to how ball lightning is made, stating how his coils produced two waves which interlocked, and it's the resulting sphere (ball) of energy that produces the electrostatic field.

If true, then there would appear to be no limit to the size and power of artificial ball lightning, because the Aether wave that produces it can be customised to any dimension or strength.

Disregarding its offensive military potential, artificial ball lightning could be projected to any location for defence against incoming enemy aircraft, missiles or ships. Also, it could be used to protect a town or city from enemy fire, by placing a dome (hemisphere) over itself and vaporizing any object that touches it.

Other applications would include protection of sensitive targets from natural or manmade disasters like storms, fires, pollution, or even incoming asteroids.

MATERIALIZATION AND DEMATERIALIZATION

Purely Theoretical

"According to the adopted theory, first clearly formulated by Lord Kelvin, all matter is composed of a primary substance of inconceivable tenuity, vaguely designated by the word 'ether'... By being set in movement, ether becomes matter perceptible to our senses; the movement arrested, the primary substance reverts to its normal state and becomes imperceptible... Thus, by the help of a refrigeration machine or other means for arresting ether movement and an electrical or other force of great intensity for forming ether whirls, it appears possible for man to annihilate or to create at his will all we are able to perceive by our tactile sense...Could he do this, man would then have god-like power, for he could create any kind of material substance, of any

size and shape, seemingly out of nothing, and he could make all perceptible substance revert to its primary form, lose itself forever in the universe."

–Nikola Tesla
Published Letter,
19th April, 1908

The above quote relates to an alternative view of the inner workings of modern Quantum Theory.

As mentioned throughout his career, Tesla believed atoms and their particles were a product of the Aether, and thus he did not believe space was empty. This concept was nothing new, many Victorian-era scientists already believed the notion of matter creation from the Aether. Even Dmitri Mendeleev, the Russian scientist who formulated our modern Table of Elements, included "Zero Group" gases below the Noble Gases. Later on, he attributed these near massless gas particles to the Aether. So Tesla, Mendeleev, Maxwell, Lodge, Faraday, Kelvin and many other great scientists of the day believed the Aether was the foundation for physical substance.

As the previous quote indicates, men like Tesla and Kelvin believed that materialization and dematerialization may have been a possibility, and manipulation of the Aether was the key to creating and annihilating real matter. What this could suggest, is that the all-pervading field is infused with a highly complex algorithm which helps give atoms and particles their characteristics.

If this process is possible, then it does make one wonder, could materialization bring a whole new meaning to the term 3D Printing.

CHANGING THE ELECTRICAL POTENTIAL

"With short waves, I readily attained activities of 20,000,000 horse-power. I found that when the waves were of a certain length it was dangerous to operate machines, because in the plants and buildings, as far as ten and twelve miles away, sparks flew from all conducting objects. Sometimes this ability was so intense that a shower of sparks could be seen. I also observed that sparks were produced on dry sand which were clearly visible at night. One may readily perceive how, under such conditions, powder and similar substances would be ignited."

–Nikola Tesla
Reported in the Wanganui Chronicle,
15th January, 1915

Just by positively charging up an object, or negatively charging it down, Tesla was able to change its electric potential. For example, if an object was positively charged up, it would start to shoot off strange electrical phenomena like sprites, lightning or electroluminescence. One bizarre discovery he made, was when he saw metre long streamers extend off his apparatus, that were 'cool misty white', harmless to the touch and had an unnatural looking colouration. If those discharges had been of the regular electrical variety, they would have killed him instantly.

Discounting its obvious military use, which would include a way of destroying an enemy's munitions at a distance, one interesting civilian application would be its ability to shield material from electromagnetic radiation, thus creating a room-temperature superconductor.

POWER THROUGH SPACE

"I have devoted much of my time during the year past" he said, *"to the perfecting of a new small and compact apparatus by which energy in considerable amounts can now be flashed through interstellar space to any distance without the slightest dispersion."* This discovery, he said, would be remembered when all his other achievements were covered with dust.

The system and apparatus embodied more than three dozen of his inventions, and with it he could transmit energy 100 miles as certainly as he could transmit energy 1,000,000 miles upwards. The energy would be different from the usual kind, however, as it would travel through a channel of less than one-half of a millionth of a centimetre."

–Nikola Tesla
Reported in the Press,
12th August, 1937

Through the use of a longitudinal *impulse current*, Tesla states how his apparatus could project Aether power out into interstellar space. The inventor believed distance was no problem, as long as an object didn't get in between the wave and its destination. From the previous description, this apparatus and the inventor's legendary Death Ray, may have operated under similar principles.

The benefits of 'interstellar power transmission' would be incredible. One example is the convenience it would provide anyone who wanted to explore outer space or the surface of another planet.

CHANGING THE COLOUR SPECTRUM

> *"This is the description of the apparatus given by one who recently visited it. From a stout beam in the centre of the rough-hewn ceiling hung three dazzling, pulsating clots of purple violet light. The room glowed with the warmth of colour. The three centres of light sent out wave after wave of a strange unearthly rich colour—a hue that is not listed in the spectrum."*

> *–Nikola Tesla*
> ***Reported in the North Otago Times,***
> ***16th October, 1901***

The newspaper article states that the inventor had found a way of creating strange new colourations which were not listed in the colour spectrum. They were 'intensely alive', and more beautiful than the natural colours we experience daily. Could this imply that there is an Aetheric spectrum and not just the electromagnetic version we are all familiar with?

WIRELESS LIGHT

"One of the many results to be achieved by the discovery, he said will be the illumination of the entire ocean at night. He has already completed details for the erection of a plant at the Azores for the lighting of the ocean at night, so that such a disaster as that which befell the Titanic could not be repeated."

–Nikola Tesla
Reported in the Poverty Bay Herald,
21st January, 1915

Numerous photographs show Tesla holding a glowing wireless bulb in his hand, while a resonant coil provides the power. But there was another lighting system he developed that didn't need a bulb at all. In the above article, it describes how the bulb-less lighting system could be used at night to illuminate a stretch of the sky over the ocean, so ships could see each other as well as icebergs.

Alternatively, technology like this could theoretically be used for lighting city streets at night, as well as illuminate tunnels, cave systems and even underwater. And because it doesn't require bulbs or electrical wiring, the operator could quite easily change the colour or light intensity.

Lastly, even though the following diverts from the subject above, there is no particular reason why similar principles couldn't be used for efficiently raising and lowering the temperatures in both air and water.

IMPULSE TELEVISION

"… perfect television, Tesla's system differs from all other present and proposed systems in that it does not employ Hertzian or similar free radiations. It is purely and simply a system of one-wire conduction—eliminating the wire and substituting the earth in its place."

–Nikola Tesla
Reported in the Press,
10th September, 1927

Before John Logie Baird and Philo Farnsworth developed our modern day television system in the 1920's, Tesla was already refining a method for using his Aether currents to project High Definition signals to any distance without interference or decay. Tesla states how his system had been worked out in all facets before 1917.

While Tesla never spoke of holograms, one interesting property *impulse currents* possess, is they can project whole 3D images to any point, a process that would be similar to projecting artificial ball lightning. But stranger still, because *impulse currents* are able to project force, they could theoretically make the image feel like it was solid.

GEODYNAMIC RADAR

"Stationary waves in the earth mean something more than mere telegraphy without wires to any distance. They will enable us to attain many important specific results impossible otherwise. For instance, by their use we may produce at will, from a sending-station, an 'electrical–effect' in any particular region of the globe; we may determine the 'relative position' or 'course' of a moving object, such as a vessel at sea, the 'speed'; or we may send over the earth a wave of electricity travelling at any rate we desire, from the pace of a turtle up to lightning speed."

–Nikola Tesla
Reported in the Evening Star,
26th September, 1900

While we use radar and sonar detection today, various statements made by Tesla indicate he was also developing a system for detecting objects in the air and ocean. From what the inventor said, the system would use his highly penetrative currents to see through fog, cloud, water, dust, and even the earth itself. Tesla declared many times that the currents he was dealing with were extremely strong, making it impossible to hide. The inventor presented the radar system to the US war department but was turned down. Interestingly, one of the board members who oversaw Tesla's proposal was none other than Thomas Alva Edison.

While *impulse currents* are able to penetrate through objects, the inventor states they could also be tuned to reflect off different elements. That would allow someone like a geologist to make accurate readings of various minerals, as well as oil and gas deposits deep within the earth's crust.

BIO AND PSYCHODYNAMIC ASPECTS OF WARDENCLYFFE

"By means of this machine I may pass at least half a million volts of electricity through a man without injury. Indeed, it may be utilised with beneficial effect.

I am of the opinion that electricity in this way may be used as a means of stamping out internal disease."

–Nikola Tesla
Reported in the New Zealand Herald,
20th October, 1900

When author John J. O'Neill described Tesla's health-enhancing platform, and how it was going to change the world, he was only giving us part of the story. Tesla often spoke about a method by which he could cure many diseases, as well as build up the immune system to bring vitality and well-being upon an individual. Some researchers have suggested how Wardenclyffe wasn't only going to send power and signals, but was also going to dispatch health-enhancing currents straight into a person's body. This would help an individual de-stress and fight off disease. Now, whether people morally agree or disagree with this aspect of Tesla's radio system is another issue, but as for the inventor, all he wanted to do was enhance the human condition, not make people into zombies.

There is much more to Tesla's health-enhancing discoveries than what is mentioned here, so it is a subject that needs to be presented along with the works of others, so an adequate argument can be submitted.

BENEFITS FOR HUMANITY

Because Tesla was such a prolific inventor, it's hard to get a grasp of the full extent of his achievements. Some have suggested his Aetheric discoveries would have affected every area of our lives, and no field of exploration would be left untouched. If that is the case, and the technology was used to improve our lives, then the freedom and independence attained by people would be unlike anything seen before.

So how would Aether technology better the lives of individuals as well as society as a whole? Firstly. by using wireless power we wouldn't need to purchase electricity or fuel for our houses and vehicles, and because the energy is unlimited, we could run appliances like air-conditioning and heating indefinitely. Also, it wouldn't matter where you lived, whether it was in the depths of Siberia, Amazonia or the Sahara, you would have the same access to power as someone who lives in New York, London or Tokyo.

It doesn't stop there though, the flow on effect of free wireless power would benefit all areas of our lives, from food production to raw materials and even the manufacture of complex machines that are currently expensive to make. By cancelling the energy bill, those commodities would plummet in price, leading to a boon for the consumer.

Another interesting benefit Aether technology could give us, is in the eradication of many diseases which plague our crops. Unlike genetic engineering, it would be a totally harmless process, and if it was used along with artificial ground heating and lighting, it would have a profound effect upon the world's crop production.

Those same benefits would also extend to the availability of pure water which requires expensive infrastructure to transport. By using free power to purify dirty or salty water, and even condense water vapour from the atmosphere (a process called 'vapour condensing'), we'd free ourselves from many of the restrictions countries face when trying to source good drinking water.

What about the environment? The potential rewards the new technology could bring us are staggering. We would finally have the ability to clean our natural surroundings from many of the pollutants that harm it, including natural phenomenon like oil seepage.

Current technologies that would become obsolete are combustion engines, household fires, oil/coal/gas-fired power plants and nuclear installations, thereby making smog and water pollution a thing of the past. Some of the spin-offs from the changeover are certain to save billions of dollars in medical and disability costs associated with pollution. Furthermore, governments would not need to spend billions of dollars cleaning many of the chemical disasters that sometimes occur in our oceans.

In its biodynamic format, the technology could be used to help protect delicate ecosystems from marauding pests. By using a tuned *impulse current* to interfere with the electrical system of a specific plant or animal, an organism could be killed without any detrimental effect to other animals or plants around it.

Lastly, if mankind converted across to Aether technologies, the benefits for the world would be incalculable. The potential to free humanity from oil dependency and food shortages, would lift billions of people from the scourge of third world poverty.

So, could this incredible technology deliver on its perceived promises, and if so, then why is it not available to us? There does appear to be multiple answers to that question. While mainstream science has rejected the Aether and the technologies Tesla was working on, highly secretive groups apparently haven't. But can a claim like this be backed up with any proof? The answer I believe is 'yes'. There seems to be evidence pointing to the fact that we have already entered into a frightful new era of warfare.

The trail of evidence starts with how we interpret physical processes. For example, after all the time and money spent trying to understand the quantum world, we still don't know how atoms and particles are made. Could some of the quantum mysteries that have plagued modern science for decades, be solved by reintroducing an all-pervading field?

In the next chapter, we take a look at the origins of modern physics, and why we in the twenty-first century may be mistaken in believing in a universe without the Aether.

118

CHAPTER EIGHT

A HOUSE DIVIDED

"If you are not completely confused by quantum mechanics, you do not understand it."

–John Wheeler

THE PILLARS OF modern physics have been with us for a little over a century now, and the scientific community believes it has a pretty good handle on how the universe works. So when it comes to the world of the super small, and their belief that atoms and particles play the central role in Quantum Mechanics, it becomes clear there is no need for a nineteenth-century Aether.

But what if the information they have gathered has been misinterpreted, and if so, what else could it tell us?

It is not my intention to go through every discovery and try to prove which one is right or wrong. Rather, it is to see whether there is any evidence for some kind of Unified Field Theory (UFT) within Quantum physics. While the UFT is a term accredited to Einstein, the concept of an all-pervading field that includes gravity has captivated scientists for hundreds of years. Its importance cannot be underestimated, because it wouldn't just unify all of the scientific disciplines, but it could also provide scientists with a way of engineering everything from particles to planets.

In this context, Quantum is a Latin word that simply means, 'a discrete quantity of energy', or it could be said, a little piece of something. And when it comes to the atomic world, the first thing that needs to be established is whether atoms are materialistic (think billiard ball in space), or whether they are some kind of field manifesting out of a larger field (think whirlpool on the ocean).

COPENHAGEN INTERPRETATION

So, what is the Copenhagen Interpretation? It is an honorary title given to one of the more popular theoretical interpretations of Quantum Theory. First presented by Niels Bohr in Copenhagen, Denmark, it states that when an observer looks at a particle from a distance it will appear as a solid object, but when they try to close in, it begins to act like a wave. One way to comprehend this strange enigma, is to imagine a particle as a mirage on the horizon of a desert. At a distance, the mirage appears as an object, but as you get closer to it, the mirage collapses, and all you are left with is shimmering waves.

When it was first discovered, it shook the scientific community to its core. How could it be both a particle and a wave at the same time?

Alongside Bohr's groundbreaking work, another physicist by the name of Werner Heisenberg, was also developing a theory. During the 1920's Heisenberg worked with Bohr in Copenhagen, and it was there that he came up with his now famous theory on particles. It was called the Heisenberg Uncertainty Principle (HUP), and in 1932, he was awarded the Nobel prize for physics.

In simple terms, HUP implies that the position and momentum of a particle cannot be truly known when measured at the same time. In the real world, this amounts to absolute idiocy. For example, we know a cars position and velocity just by measuring it, but in the quantum world, those details are more uncertain, or what is termed Probabilistic. The subject of Probabilities is far more complex than what is written here, but what it essentially says is that particles operate in a fuzzy world where our rules don't apply. Ever since its introduction, HUP has annoyed scientists no end and has even led some to question the very foundations of reality itself.

By the end of the 1920's, Werner Heisenberg and Niels Bohr's theories on particle/wave duality were pretty much cemented into mainstream science. But how does this relate to anything in the real world? In the scheme of things, it is quite important because it will affect the direction of where particle physics will head for the next century.

PARTICLE-WAVE DUALITY

The particle-wave concept is much older than Niels Bohr and Werner Heisenberg, in fact, it goes back about three hundred and fifty years, to the time when scientists began studying the nature of light.

In the late seventeenth century, a Dutchman by the name of Christiaan Huygens (pronounced Hoygens), proposed that light might be a wave. According to his theory, light waves propagated longitudinally through a stationary (static) Aether. Conversely, Englishman Sir Isaac Newton believed light rays consisted of a stream of particles that were accompanied by vibrations in the Aether.

Newton's particle theory held sway for many years, but it was becoming apparent light did indeed possess wave-like qualities. Even though both theories used an Aether, it was Huygens wave-only theory, that needed the all-pervading field like watery waves need an ocean. The only problem was, the Aether appeared to be almost impossible to detect.

The theory of an all-pervading field goes back to antiquity, but with more recent research from people like Descartes, Newton, and Huygens, the Aether model has grown in scope and complexity.

Even the great Scottish physicist, James Clerk Maxwell, invoked an Aether type medium when he mathematically unified magnetism and electricity, outlining his thoughts in *A Treatise on Electricity and Magnetism Vol II*.

> *"Whenever energy is transmitted from one body (particle) to another in time, there must be a medium or substance in which the energy exists after it leaves one body and before it reaches the other."*

–*James Clark Maxwell*

Maxwell was a supporter of a medium (Aether) throughout space, and that it carried electric and magnetic waves. For him, there was no such thing as an absolute vacuum; he even attacked the scientific community who didn't support his view, telling them they believed in a type of pseudo-religion called Scientism.

Another prominent scientist mentioned in the previous chapter also believed in an all-pervading field. Dimitri Mendeleev proposed that the Aether was made up of gaseous elements of very low atomic weight, so he placed them under the Nobel Gas group of his Periodic Table.

Prior to the twentieth century, there were many attempts at trying to detect the Aether. Some, like James Bradley, believed it was a 'static field'. This is analogous to believing the earth moves through the medium like a submarine moves through a still ocean. Yet others claimed the Aether was a 'dynamic field', which implies that bodies like the earth were, in fact, dragging a piece of Aether along with it, just like a magnet drags iron filings around on a piece of paper. Nikola Tesla was in the second camp, his belief was that the Aether was like a gas, and the all-pervading field and electricity acted in a similar fashion.

After much debate, it all came to a head in 1887, when Albert A. Michelson and Edward W. Morley conducted a test which became known as the Michelson-Morley Experiment. In essence, a device called an interferometer sent two beams of light in two different directions, which were then measured as they moved through the Aether. One beam moved directly into the Aether as the earth moved through it, while the other beam did not. If the two beams returned home at different times, the Aether existed. To the surprise of many, the conclusion of their experiment came back with a 'null' result, so it was assumed there was no Aether, or more precisely, no 'static Aether'.

However, the Michelson-Morley result did not quench the fervour of the Aetherists, with the battle raging on for several more years. Even Albert Michelson himself stated that they did 'not' not detect the Aether; what they did do was find a way of not detecting it.

But as new particle-only theories gained acceptance, it was becoming harder for Aether proponents to keep their theory relevant.

Of course, the situation wasn't helped by the introduction of Einstein's two famous theories, which eventually saw the Aether replaced by a non-physical space. In Special Relativity (1905), light in space is placed in an absolute vacuum, and must always be a constant 186,000 miles per second (300,000 kilometers per second).

Also, in General Relativity (1915), a physical/engineerable space medium (Aether) responsible for gravity, is replaced by a four-dimensional, non-physical, non-engineerable entity called Space/Time. It must be mentioned, that Tesla totally disagreed with Einstein when it came to Relativity, stating "How can something [the earth] act upon nothing [empty space]" when it came to gravity.

So what does this all mean for Aether theory? Regardless of what the mainstream scientific community knows about Quantum Theory, and the so-called emptiness of space, there are still many mysteries that appear to defy any rational explanation.

To mainstream science, those mysteries are truly baffling, but to Aetherists, they might just reopen the door to an all-pervading field. The following six examples describe some of those enigmas, with the first of them bringing into question the so-called universal speed limit.

QUANTUM TUNNELLING

There are many strange things that happen in the quantum world, but one that is truly bizarre is where particles perform what is known as Quantum Tunnelling.

First observed by German physicist Friedrich Hund in 1927, and independently confirmed by Leonid Mandelstam and Mikhail Leontovich. It is a discovery which has grown in complexity and scope over the years, and according to an article in the *New York Times*, Dr Raymond Y. Chiao of Berkley states why scientists are still bewildered by the tunnelling process.

"We find" Dr Chiao, "that a barrier placed in the path of a tunnelling particle does not slow it down. In fact, we detect particles on the other side of the barrier that have made the trip in less time than it would take the particle to traverse an equal distance without a barrier—in other words, the tunnelling speed apparently greatly exceeds the speed of light. Moreover, if you increase the thickness of the barrier the tunnelling speed increases, as high as you please. "This is another great mystery of quantum mechanics."

–New York Times,
22 July, 1997

What this suggests is, that when a stream of particles encounters a barrier, most, but not all, will be stopped. The few that do go through however, are believed to traverse the distance at speeds greater than light. To get around this problem, scientists believe Probabilistic Distribution (HUP) is the reason the particles move as they do. In other words, the fuzziness of HUP allows the particles to be in multiple positions at the same time.

While Quantum Tunnelling is real, it must be asked, could this phenomenon be as simple as Probabilities? Also, if they are moving through the barrier at speeds beyond that of light, one must wonder, if there is a more fundamental field helping the particles achieve this feat?

CASIMIR EFFECT

On the one hand, General Relativity states how space is empty, while Quantum Theory believes something might be there. Is space empty, or is it full? From a discovery made just after the second world war, it might appear space is actually full.

In 1948, Dutch physicist Hendrick Casimir theorized, then proved, that if you took two uncharged conductive plates, and brought them extremely close together in a vacuum, a mysterious attracting force appeared between them. Since then scientists have proposed various theories, with the most popular advocating the presence of virtual-photons, or little entities which pop in and out of existence from the active-vacuum. It is this super-fast manifestation of the virtual-photons that causes the plates to pull together.

To some, the previous description might imply that

an Aether type field is operating throughout space. But that isn't the case, what scientists actually believe is, some kind of electromagnetic soup is fluctuating throughout the universe.

But is this correct? It is known that a virtual-photons life-cycle is ruled by HUP, so their short lives cannot be directly observed. So it has to be said, whatever is causing the Casimir Effect is still up for debate.

There are some though, who believe the Casimir Effect proves that so-called empty space is anything but empty, and also how it might indeed be engineerable, even if only on a small scale.

BOSE-EINSTEIN CONDENSATE

During the 1920's, Satyendra Bose and Albert Einstein co-wrote a theoretical paper describing how certain particles behaved when their temperatures were lowered to almost zero degrees Kelvin, or -273.15 degrees Celsius (-459.67 degrees Fahrenheit).

Called the Bose-Einstein-Condensate (BEC), the experiment uses Bosonic particles, that are cooled until they turn into a condensate. When the super-cooling is achieved, strange things begin to happen. For instance, particles will lose their separate identities and act like they are one big super-particle. Furthermore, certain types of super-particles exhibit new characteristics, like super-conductivity (no resistance), and superfluidity (liquid flowing uphill). Stranger still, scientists have found a way of retarding or stopping the flow of light within a BEC, but still have no definitive answer as to why it happens.

This raises an interesting question, where do the laws come from that govern the BEC, are they in the individual particles, the super-particles, or do they come from space (Aether) itself?

It is interesting to note what eminent physicist David Bohm theorised about the quantum world. He believed elementary particles had extremely complicated internal structures and that they acted as amplifiers of information contained in a quantum wave. His theory is called the Implicate Order, and it uses some kind of all-encompassing information realm to connect all the particles in the Universe. Could Bohm's theory be correct? If so, then it would explain strange particle behaviour such as faster than light communications between entangled particles (known as Quantum Entanglement).

Whatever the final outcome is for this mysterious discovery, one thing is certain, the quantum world is much stranger then we could ever have imagined.

EMPTINESS OF THE ATOM

When we touch an object, we perceive it to be tangible and solid, but the same cannot be said for atoms and their particles. In that world, nothing is what it seems.

To comprehend what we are dealing with, we must first understand the sizes involved. For example, atoms are so minute that there are more of them in a teardrop than there are stars in the observable universe. Also, their inner dimensions are so small, that if the nucleus was the size of a basketball, then the outer edge (electron shell) would be about seven miles (11.2kms) away.

So why are these details so important when it comes to understanding atoms and their particles? Because recent discoveries have revealed they are for all intents and purposes, empty.

What scientists have found is, the atom is more than 99.999999999% empty space. Worse still, they found that sub-atomic particles like Electrons, Protons and Neutrons, are themselves practically empty. This has led physicists to search for even smaller particles called Quarks, a particle they believe gives mass to the nucleus. Interestingly, some computer models have suggested that Quarks are made up of fluctuations from the active-vacuum.

So, the perception that atoms and particles are solid objects like billiard balls, is simply untenable. According to some scientists, it's possible that energy from the active-vacuum is responsible for inflating and sustaining atoms and particles, but because of Heisenberg's Uncertainty Principle, it is nearly impossible to directly detect.

But just as it was mentioned earlier in the chapter, the mainstream scientific community does not believe the active-vacuum has any relationship with the nineteenth century Aether.

HYDROGEN GROUND STATE

The classical view of the atom developed during the early part of the twentieth century, states that a single particle called an electron, orbits around the nucleus of the Hydrogen atom, like a planet orbits around the sun.

The only problem is, the electron continuously radiates EM energy as it orbits, which should cause the particle to spiral down into the nucleus until the atom collapses. To get around this problem, most scientists have adopted Probabilistic Distribution (HUP) for the electron. In essence, they took the electron from a particle and made it into a wave, thus circumventing the energy loss problem.

But is that entirely correct? According to some scientists like Hal Puthoff, the electron is, in fact, absorbing energy from the Zero-Point-Energy (ZPE) field which permeates throughout space. It is this interaction between the ZPE and the electron, that allows the particle to periodically readjust its orbit away from the nucleus.

Likewise Nikola Tesla believed atoms were receiving their energy from an unseen field, and through his experiments, he confidently maintained it could only have been the Aether.

So, if the electron is absorbing energy from the ZPE field, and is transforming it into light, then only one question remains, what really is Zero-Point-Energy?

THE DOUBLE SLIT EXPERIMENT

The final example goes to the very heart of the particle-wave, and HUP debate; it's a simple experiment, but quite hard to describe. First performed by Thomas Young in 1801, the Double-Slit is an experiment where particles are shot at a screen with two slits in it. After the particles move through the slits, logic states that they should collect in two clumps on the back screen, but this is not what happens. What actually takes place is, a strange interference pattern appears on the back screen.

If that wasn't weird enough, if you shoot a single photon at the double slits, it will appear to move through both slits at the same time. Also, if scientists put detectors in a certain spot and try to observe the particles, the interference patterns disappear.

There is far more to this experiment than what has been explained above, but it's enough to raise serious questions about what might be happening in and around particles. So why do particles change their state when we try to observe them — could an unseen field be causing some kind of interference?

BACK TO THE FUTURE

The previous six examples reveal some of the bizarre behaviour taking place in the quantum world, and it doesn't take long to see that there are big gaps in our understanding of how atoms and their particles work. But what about our techniques for observing them, could they be a part of the problem, if they are, then that could change everything?

Take, for example, Werner Heisenberg's Uncertainty Principle, it has been around for almost a century, and even though it has been one of the bedrocks of modern quantum theory, signs are starting to appear that it might be misinterpreting what is really taking place around particles.

Recent discoveries by scientists in Toronto, Canada, may have opened the door to a better way of understanding particles. In their experiments, they were able to take a series of really small measurements instead of one big one. By doing that, they did not disturb the particles of light, which allowed them to get a more accurate picture of the test subject. While there is still much to learn from the preliminary Canadian experiments, initial tests suggest that for the last century we may have been trying to observe quantum behaviour with a sledgehammer, instead of a key. What this could mean is, that one day we might be able to understand a particles position and momentum with absolute certainty. If true, then surely Heisenberg's Uncertainty Principle as we understand it, is wrong.

Even today in the early twenty-first century, some physicists still have trouble accepting quantum theory as it is taught. They believe it is hard to grasp, in some ways leaves more questions than it answers, and will probably never bring a Unified Field Theory which includes gravity. It does have to be said though that mainstream science does accept the notion that space is not empty, and even argues how Quantum Entanglement (HUP) is responsible for how Quantum Computers work. But that does not refute the argument HUP might be wrong when it comes to how particles behave at the micro level.

If a particle is capable of moving through walls, changing its state, performing faster than light communications, or transforming invisible space energy into visible EM energy, then aren't particles acting in a way which is similar to how Tesla described the Aether? The Serbian states more than once that Aether energy could penetrate matter, transmute particles, move at any speed, and was the power source for particles and atoms. Tesla, like many scientists at the time, believed a Highly-Dynamic, Super-Complex, All-Pervading field of sound (the Aether) was the agent responsible for sustaining particles and atoms. So just may be the answer to the weird quantum behaviour seen in the micro world has been with us all along.

Many names have been used for this all-encompassing field. Some of them include Aether (ether), Quintessence, Plenum, Dark Fluid, Sea of Energy, Medium, and All-Pervading Field. Conversely, because of how scientists define them, it does not include terms such as Active-Vacuum, Zero Point Energy, or Higgs Field. For a start, these don't include gravity so they couldn't possibly be called all-encompassing fields. What about String Theory? While some scientists believe it has the ability to unify Quantum Mechanics and Relativity, pointing to its elegant mathematical equations as proof. Other scientists are far from convinced, they think some of the concepts within the theory are fanciful and couldn't possibly work in the real world.

While scientists and philosophers still argue over the presence of an unseen medium, another source of information might help corroborate the Aethers existence. It is a source that has been with us for thousands of years, and has inspired some of our greatest scientists like Nikola Tesla, Sir Isaac Newton, Lord Kelvin, James Clerk Maxwell, and Johannes Kepler. That source is the Bible, or more precisely, scriptures describing the unseen nature of the universe. By using the scriptures to enhance our understanding, it may be possible to bridge the gap separating the visible and invisible universe.

CHAPTER NINE

ROSETTA

"The gift of mental power comes from God, Divine being, and if we concentrate our mind on that truth, we become in tune with this great power. My mother had taught me to seek all truth in the Bible."

–Nikola Tesla

Disclaimer

In *A House Divided*, (chapter 8), it asserts how there are still issues within modern quantum theory, and they could possibly be solved by reintroducing the Aether. In this chapter, the idea is to provide an alternative approach to explaining how the all-pervading field might interact with atoms and their particles. But to do that, the Aether field must be elevated into the primary position, while the atom and its particles are relegated to secondary.

After each Biblical quote is presented, the author will include a demonstrable experiment to help the reader understand what is being proposed, and from that, show why space itself might be a dynamic field, capable of creating our physical reality.

~

"By the word of the Lord the heavens were made, and all the host of them by the breath of His mouth."

–Psalms, Ch.33:6

One of the biggest conflicts between Faith and Science is the perception that somehow scientific reason has disproved the Bible, but as we have seen, it is quite the opposite: the deeper scientists dig, the more we find out that Biblical writers may have had the concepts right all along. The more science discovers about the quantum world, the more frequently God is invoked. Could we have side-lined Biblical truth based on flawed scientific understanding? What if the Aether is the nexus between the material world and the spiritual?

This chapter is not implying the Aether is God, it is merely suggesting how the Aether is a part of His creation, and that its presence answers some of the biggest mysteries facing science, and in particular, quantum physics.

If the God of the Bible is the Creator, then shouldn't some of His secrets be found in His word? The truth is, we do find clear references to various secrets, with some of them now understood by science, while others still remain unknown. Four examples will be presented.

- In Isaiah, chapter 40:22, it states how the earth is circular – *"He sits enthroned above the circle of the earth..."* – disproving the misconception that everybody thought the world was flat.
- In Job, chapter 26:7, we are told that the Earth is suspended in space, or in other words, on nothing: *"He spreads out the northern skies over empty space; he suspends the earth over nothing."*
- Also mentioned in the Book of Job, are the two mysterious creatures called *Leviathan* and *Behemoth.* From their description, they would appear to be prehistoric, but in reality, must have been present at the time of the Old Testament. It is interesting to note that, in recent years, paleontologists like PHD, Mary Schweitzer, have found proteins and soft tissue in T Rex, Hadrosaur, Triceratops, and Seismosaur (candidate for *Behemoth*) fossils, which should make their alleged sixty million year age impossible. According to current understanding, most scientists agree that even in a best-case scenario, non-fossilized bio-material should not last more than a hundred thousand years, which is a far cry from the sixty million years for the animals listed above.
- In Deuteronomy, chapter 28:12, it states how there are vast store houses from which the rains come from: *"The Lord will open the heavens, the storehouse of His bounty, to send rain on your land in season and to bless all the works of your hands..."* It's now known by science, that there are rivers of water vapour larger in volume than the Amazon, moving through the earth's atmosphere.

Also, there are the storehouses of the deep. For two centuries, scientists and schoolteachers have gently mocked the Bible's account of a Great Flood, pointing out there is simply not enough water on Earth to double the depth of the oceans. The mocking stopped in 2014 when scientists discovered vast springs on the ocean floor pumping water through a planetary circulation system: In *New Scientist*, 12th June 2014, this new discovery is described.

> *"A reservoir of water three times the volume of all the world's oceans has been discovered deep beneath the Earth's surface... The water is hidden inside a blue rock called ringwoodite that lies 700 kilometres* [435 miles] *underground in the mantle... The hidden water* [through oceanic springs] *could also act as a buffer for the oceans on the surface..."*
>
> **–New Scientist,**
> **12th June, 2014**

While estimates vary, it appears there could be enough water on Earth to cover Everest three times over. Those approximations include H_2O in the air, on the surface, and deep underground.

In the Book of Job, 38:16, written thousands of years ago, a farmer is given a secret that science did not discover until 2014: *"Have you journeyed to the springs of the sea or walked in the recesses of the deep?"*

In Genesis, chapter 7:11, the Bible explicitly states how the flood was caused:

"...in the second month, on the seventeenth day of the month, on the same day all the fountains of the great deep burst open, and the floodgates of the sky were opened."

And in Genesis, chapter 8:2, the flood ended: *"Also the fountains of the deep and the floodgates of the sky were closed, and the rain from the sky was restrained."*

Just as there are scriptures revealing some of the visible secrets of the universe, there are also passages describing invisible ones. These discrete passages provide insight into the world of the super small, but there is a catch: the scriptures do not appear to endorse a particle only universe (Quantum Mechanics), and likewise, they do not support an empty space (Relativity). What the scriptures do suggest is that sound is used to manipulate a 'fluid medium', and that this process is responsible for producing and sustaining all of the physical matter in the universe.

The following pages present some of the Biblical scriptures concerning creation, which are accompanied by a description of how they might be linked to the Aether, as well as the atom itself.

WATERS
(Aether)

"The earth was without form and void; and darkness was on the face of the deep. And the

spirit of God was hovering over the face of the waters. And God said, Let there be light, and there was light."

–Genesis, Ch.1:2-3

According to several scriptures at the start of Genesis, chapter one, the physical universe was spoken into being by God. One of those Scriptures (mentioned above), reveals two important aspects of this process. Firstly, the *waters* are the Aether field, and secondly, *light* (EM spectrum) is a by-product of sound (super-high frequency resonance) upon that field.

This verse implies the *waters* (Aether) were created before the physical universe came into being, and that light and matter were created from it.

When it comes to the *waters*, the energy it contains is absolutely staggering, and depending on the equations you adhere to, they vary from 10^9 joules per cubic meter up to 10^{113}, that is, 10 with 113 zeros behind it. The best way to describe the density of energy, is to use a quote by physicist Richard Feynman, when he states:

"The energy in a single cubic meter of empty space is enough to boil all the oceans of the world."

–Richard Feynman

Richard Feynman references a cubic meter, while John Wheeler mentions a coffee cup. The exact figures for the density of the all-pervading field are still being debated, but the general consensus amongst many scientists is that it definitely exists, is super dense, and it most certainly interacts with atomic particles.

But just as it was mentioned in the previous chapter, scientists have a different interpretation for the sea of energy, making it into a cauldron of electromagnetic flux, instead of an all-pervading field that includes gravity.

Even Max Planck, the father of modern quantum physics, knew there was more to the theory of a particle only universe, stating his thoughts over how particles are made.

> *"All matter originates and exists only by virtue of a force which brings the particle of an atom to vibration and holds this most minute solar system of the atom together. We must assume behind this force the existence of a conscience and intelligent mind. This mind is the matrix of all matter."*
>
> **–Max Planck**

Max Planck, along with scientists like Lord Kelvin, Nikola Tesla and David Bohm, have stated how particles must be receiving energy and information from an outside source, otherwise, they would collapse. If that is the case, then just maybe the Aether is imbued with a highly complex algorithm which controls the processes of atoms and particles.

In my opinion, if the Bible is correct, then the most intriguing aspect about creation is the fact that it was spoken into being by the Lord God, and everything we see and feel in the visible universe was/is created by what would appear to be an insignificant agent called sound.

VIRTUAL FIELD
(Invisible (Aetheric) part of the Atom)

"By faith we understand that the worlds were framed by the word of God, so that the things which are seen were not made of things which are visible."

–Hebrews, Ch.11:3

In the Book of Hebrews, chapter 11, verses 3, the scripture implies that anything physical was first created through the use of sound. Tesla also fervently believed atoms were receiving their energy from an outside source. Stating: "There is no energy in matter other than that received from the environment [the Aether]" – Nikola Tesla.

So how can atoms and particles be practically empty, yet operate like they are highly complex machines? To answer this, even Quantum Mechanics has had to invoke a field it calls the active-vacuum. If atoms and particles are not solid objects, then surely, they must have an invisible, or what some might call a Virtual-Field, holding them together.

To use an analogy, it could be said that atoms and particles are likened to a whirlpool on the surface of the ocean, and that a core of bubbly froth sits at its centre. What this analogy implies is, atoms and particles are not solid objects in empty space, but are in fact dynamic fields in an all-encompassing sea of energy. Furthermore, the visible electromagnetic (bubbly froth) energy, is a by-product of the Virtual-Field.

To describe how sound might form the virtual components of atoms and particles, one interesting experiment conducted by researcher and author Stan Deyo, explains how this might be achieved.

In Deyo's experiment, (Figure 13, page 161), he uses a specific sound-wave frequency to create a stationary wave in a bottle of fluid. After the sound is applied, a messy turbulence begins to appear on the surface, then after several seconds the turbulence settles, and a sphere (or spheres) of fluid appears instantly on top. Their spin and inner wave structure keeps them separate from the rest of the fluid, and as long as the sound is applied, the sphere/s remain intact.

In that visual demonstration of Deyo's, it is easy to see why waves could be so important to atoms and their particles, and also, why it's not hard to imagine how sound in the Aether could be responsible for producing physical matter.

The following segments describe the various parts that make up the Virtual-Field.

CORNERSTONE & PILLARS
(Nucleus)

"To what were its [the Earth's] *foundations fastened? Or who laid its cornerstone."*

–Job, Ch.38:6

"The pillars of the earth are the Lord's, and he has set the world upon them."

–1 Samuel, Ch.2:8

According to the two verses above, the earth possesses two internal components that the planet relies upon, they are the *cornerstone*, and the *pillars*.

So how can a sphere have a cornerstone? It could only be one thing, the core.

Now it is assumed by modern science that the core of the earth is molten because it's still cooling from its formation billions of years ago. But if Tesla was right, and the earth is full of Aether, then it stands to reason the earth's core is not actually cooling, but is absorbing energy from the all-pervading field.

What about the atom? From the Serbian's own statements, it's clear he believed atoms received all their energy from the Aether, which suggests the atomic nucleus possesses a hidden Aetheric field. That hidden field consists of a toroidal (donut) shaped current, whose pulsing action creates EM energy.

The second component spoken of in the first Book of Samuel is a little more mysterious, and it involves what the scriptures call the *pillars*.

The scriptures are obviously not talking about physical pillars, but it could be talking about a longitudinal (tube like) Aether current, that flows true north/south through the earth.

Could the spin and changing tilt of the earth suggest that unseen *pillars* of energy are responsible for creating those effects? If the earth possesses Aetheric *pillars*, then the atom must as well. What this would mean is, the *cornerstone* and its *pillars* are the inner-skeletal structure that creates the next stage of the virtual field, the *foundations*.

Shown in Figure 9 (page 157), the longitudinal wave travels true North/South, and through the toroidal core.

FOUNDATIONS
(Outer Atom)

"Where were you when I laid the foundations of the earth? Tell me, if you have understanding. Who determined its measurements? Surely you know!

Or who stretched the line upon it? To what were its foundations fastened? Or who laid its cornerstone?"

–Job, Ch.38:4-5-6

In the Book of Job chapter 38 verses 4, 5 and 6, the scriptures mention the *foundations* of the Earth, which is describing the various layers inside the planet.

Why is it, that when many of the natural objects in the universe are cut in half, they appear uniformly layered. Stars, planets, and even life forms like plants, reveal an internal structure which looks surprisingly similar to the layers of an onion.

Interestingly, this same trait is also seen inside the quantum world. For example, it is known by science that atoms have shells, and electrons shift their orbit inwards and outwards from the nucleus, in jumps. Assuming the atom contains an internal field of Aether (Virtual-Field), then the *foundations* could be the outer-skeletal structure holding the electrons in place.

Also, if the Virtual-Field is present in particles, atoms, planets and stars, then surely there must be an internal order to the Aether itself. Could the entire Aetheric universe contain an inbuilt substructure which looks similar to magnetic 'lines of force', making this system responsible for matter, as well as forces like gravity and inertia?

The four bullet points below, along with their diagrams, describe different parts of the Aether, which includes the *foundations*, and the stationary waves that create them.

- Figure 7 (page 155), shows the 'lines of force' flowing freely throughout space.
- Figure 8 (page 156) shows what happens to the 'lines of force' when they are in the presence of matter.
- Figure 10 (page 158), shows how the *foundations* of an atom are formed by Aetheric-stationary waves, and how they are connected to the *cornerstone* and *pillars*.
- Figure 12 (page 160), shows the 'lines of force' merging between two atoms. This area is called the Virtual Corona, or in mainstream physics the Van der Waal zone

TRANSFORMATION
(Creation of Light through Sound)

"Then God said "let there be light", and there was light. And God saw the light, that it was good; and God divided the light from the darkness."

–Genesis, Ch.1:3-4

The creation of electromagnetic energy in the atom is still poorly understood by scientists. According to current theory, they believe EM energy is produced by accelerating electrons, and even though they are able to calculate what is happening during the process, they still don't fully understand why.

Maybe the Aether holds the answer to how EM energy is produced in the atom, and through its use of sound, it is capable of producing all that we see physically.

There is one interesting experiment discovered in 1934, that could give us a clue as to how this might be done. Called Sonoluminescence, it uses a high-intensity sound wave to collapse a gas bubble in a fluid. When the bubble collapses it emits a pulse of light, and under certain conditions can produce pulses of light continuously.

Here is an experiment where sound is capable of producing light, and high temperatures in a fluid. So, could an effect similar to this produce all EM phenomena from the Aether, and if so, why do we not see it happen? The answer might lie in the fact that science could be misinterpreting the processes around the electron, and that it might, in fact, be a transformer for unseen Aetheric currents.

Modern physics might have a different interpretation

of how electromagnetism comes into being, but many scientists in Tesla's day, believed the Aether was the primary agent behind atoms and their EM phenomena.

CYMATICS
(Creation of Substance through Sound)

"Then God said "Let the waters abound with an abundance of living creatures, and let birds fly above the earth across the face of the firmament of the heavens."

–Genesis, Ch.1:20

In the first chapter of Genesis, several verses describe how God spoke all living creatures into being.

So, is there a process which has the ability to produce complex shapes through sound? There is one particular area of research that could give insight into the intricate nature of sound, and it's called Cymatics.

Known about for centuries, and recently popularised by a Swiss man called Hans Jenny, Cymatics (Greek for wave) is a powerful visual tool for showing how something complex can be made from something deemed insignificant.

In the demonstration, sound from a speaker is sent into a metal plate or bowl that contains a material like sand or malt. As the material moves around, it will begin to form 2D and even 3D shapes. Then by simply flicking the dial on the audio equipment, the frequency of the sound is changed, and the material moves into a new shape.

By studying Jenny's pictures and videos, fine material can be seen taking on some of the shapes we observe in the natural world. For example, his demonstrations show spiral galaxies, hurricanes in motion, as well as geometric patterns like snowflakes, sea shells, flowers and even tortoise shells.

The beauty of Cymatics is, it has the ability to explain the various levels of existence, from the subatomic, all the way up to galactic clusters, and even the universe itself. And unlike many of today's scientific theories, Cymatics is easy to understand.

In Figure 14, (page 161), it shows the wondrous complexity sound is able to achieve, and it just might be the key to revealing how the Aether does the same.

So the concept that sound might be responsible for creation, might not be so foreign after all.

The universe with all its beauty and unfathomable complexity never fails to leave us in awe, whether it's a particular plant, animal or simply the beauty of the starry sky at night. And it's no different for those who study the world of the super small, the exquisite complexity of the invisible realm shows just how finely tuned the universe is.

If everything we see is made from a highly complex invisible field of sound, then you do have to wonder where the sound comes from in the first place. In my opinion, the sound which makes and sustains the elements is so beautifully interwoven, that it must have come from somewhere, for logic tells us, it could not have appeared from nothing.

This brings us to a statement made by one of the most influential scientists of the twentieth century, and even though some have suggested it was just a throw-away line, I would beg to differ.

"God does not play dice with the universe."

–Albert Einstein

Chance or design? Symphony or explosion? Whether it was a flippant statement or not, it quite accurately describes the complex harmonic nature of the subatomic world. This, among many other discoveries, suggest the universe could not have been created by chance, and for that reason, I give the last word to the author of the second book of Peter:

"But they deliberately forgot that long ago by God's word the heavens came into being and the earth was formed out of water and by water."

–2 Peter, Ch.3:5

Figure 1: A photographic portrait of Nikola Tesla, circa 1890.

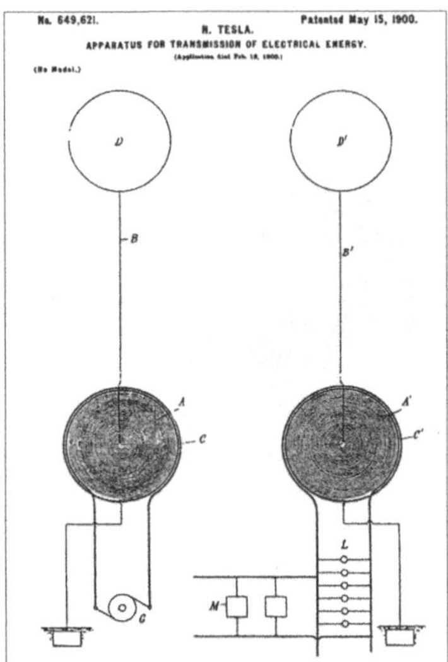

Figure 2: Granted on the 15th of May 1900, US Patent No 649, 621 describes Tesla's apparatus for transmitting electrical energy through the natural medium (Aether).

Figure 3: A photograph for *Century Magazine* (June 1900) shows an experiment containing three incandescent lamps connected to a 50-square foot wire loop. The T.M.T. is situated 100 feet from the loop and provides the resonant vibration to power up the bulbs.

Figure 4: A view of the Colorado Springs experimental station, located 6,000 feet above sea level in the shadow of the Rocky Mountains. The barn-like structure was designed to house Tesla's giant T.M.T. The building in the background is the Union Printers Home.

Figure 5: Tesla's Wardenclyffe facility on Long Island is photographed in the first decade of the twentieth century. While the laboratory looks complete the transmitting tower remains unfinished. What is unseen is the 120-foot underground antenna system.

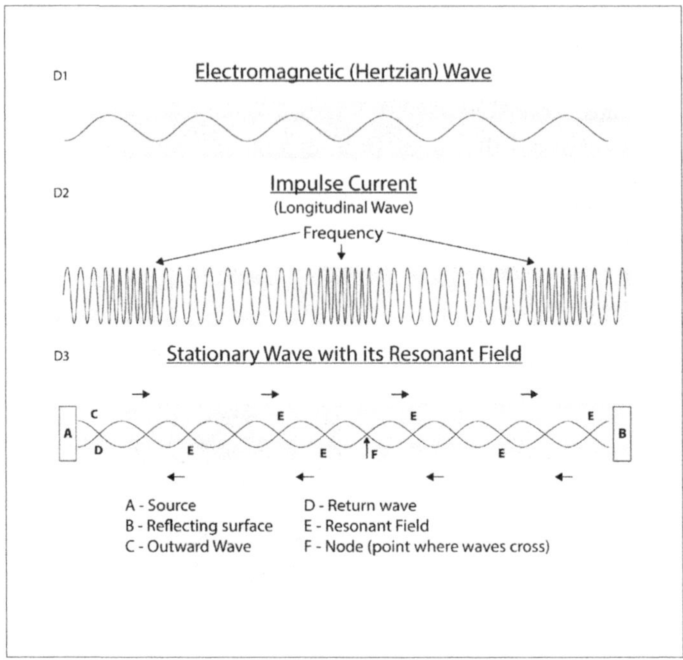

Figure 6

ELECTROMAGNETIC WAVES (D1)

Moving in a back and forth swaying motion, an electromagnetic (EM) wave is also known as a transverse or Hertzian wave. Visible light is only one small part of the electromagnetic spectrum.

IMPULSE CURRENTS (D2)

Sometimes called scalar waves, these longitudinal waves are a component of the Aether, and permeate all corners of the universe. Along its length are rarefactions (loose) and compressions (tight), this creates the frequency. NOTE: There are various types of Aether waves.

STATIONARY WAVE WITH ITS RESONANT FIELD (D3)

When a wave moves out from its source and hits a wall, it will return just like an *echo*. If the frequency is set correctly, the troughs and peaks of the two waves will align at 180° out of phase, thus creating a powerful field. One effect stationary waves produce is sympathetic resonance on an object.

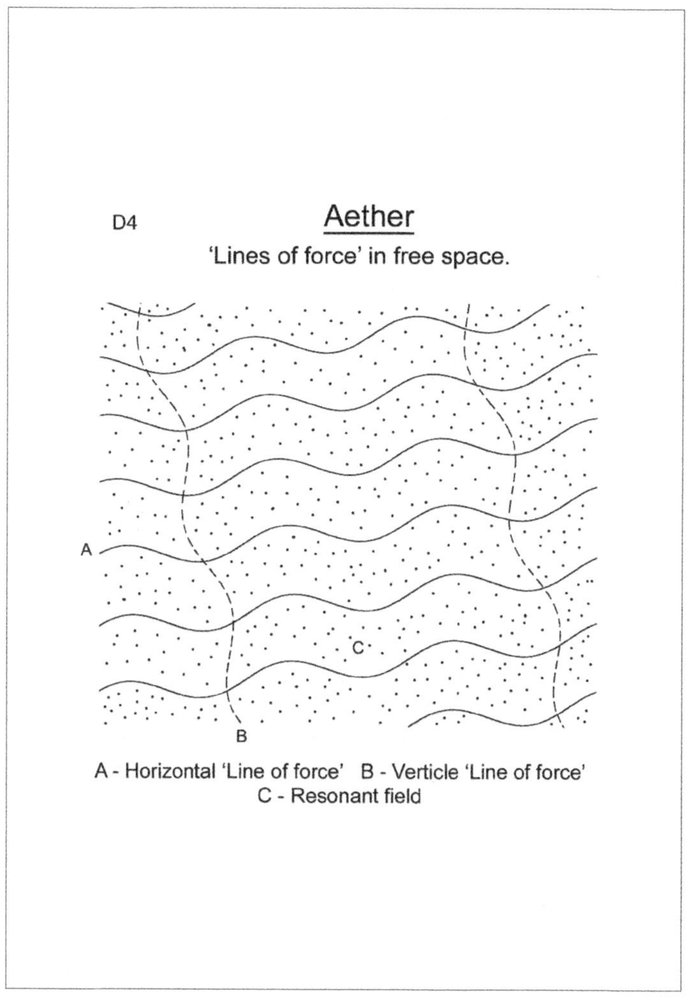

Figure 7

LINES OF FORCE IN FREE SPACE (D4)

Aetheric 'lines of force' exist and operate throughout empty space. They are not rigid or set in place, but are able to move like a spider's web swaying in the wind. All levels of existence in the universe use these 'lines of force', from the sub-atomic, all the way up to the Universe itself.

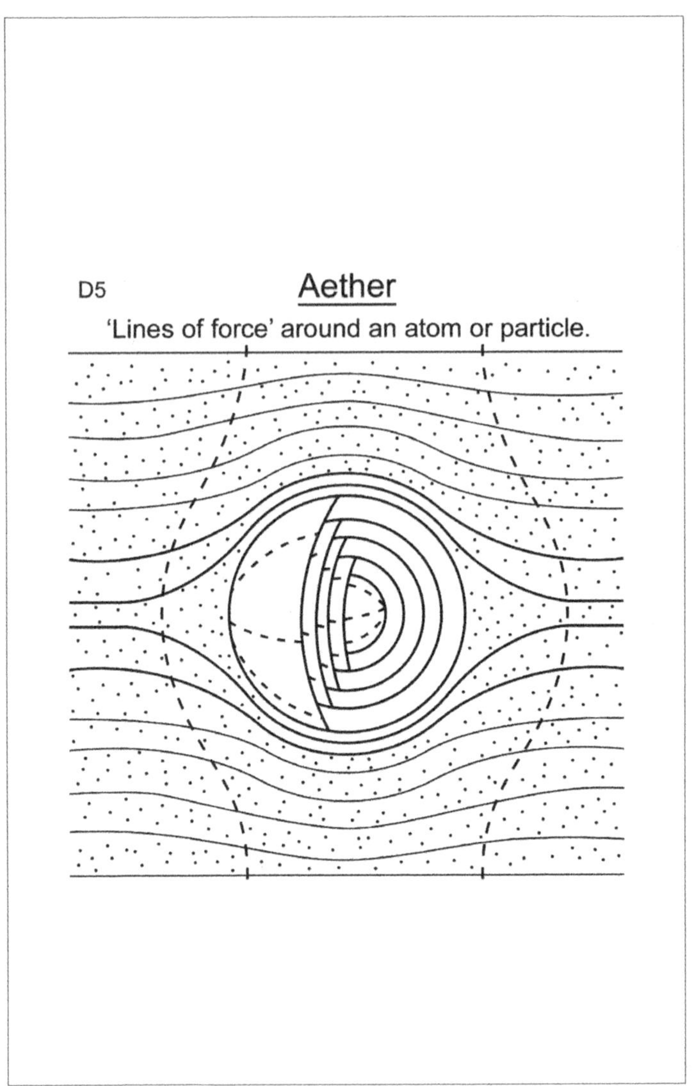

Figure 8

LINES OF FORCE AROUND AN ATOM OR PARTICLE (D5)

In the presence of matter the Aetheric 'lines of force' squeeze together. When this occurs, they impart pressure on their surroundings, causing various effects in and around the atomic structure.

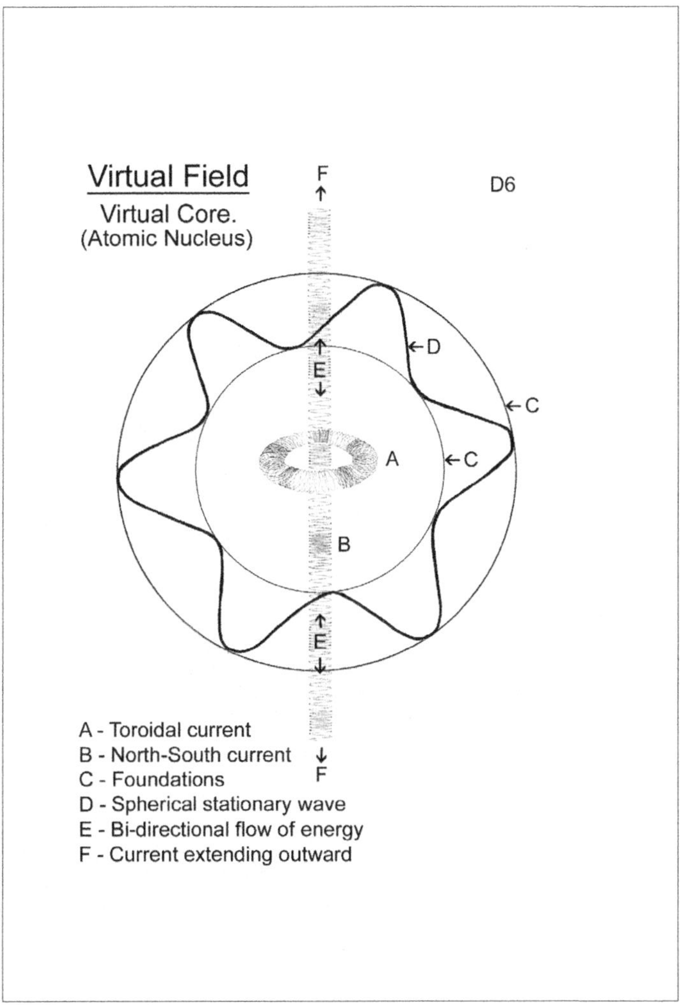

Figure 9

VIRTUAL CORE (D6)

In Aether theory, the atomic nucleus requires energy to flow in from the all-pervading field, which is then converted into electromagnetic energy. The Virtual Core (nucleus) comprises of the 'toroidal ring', and the north/south *impulse current*.

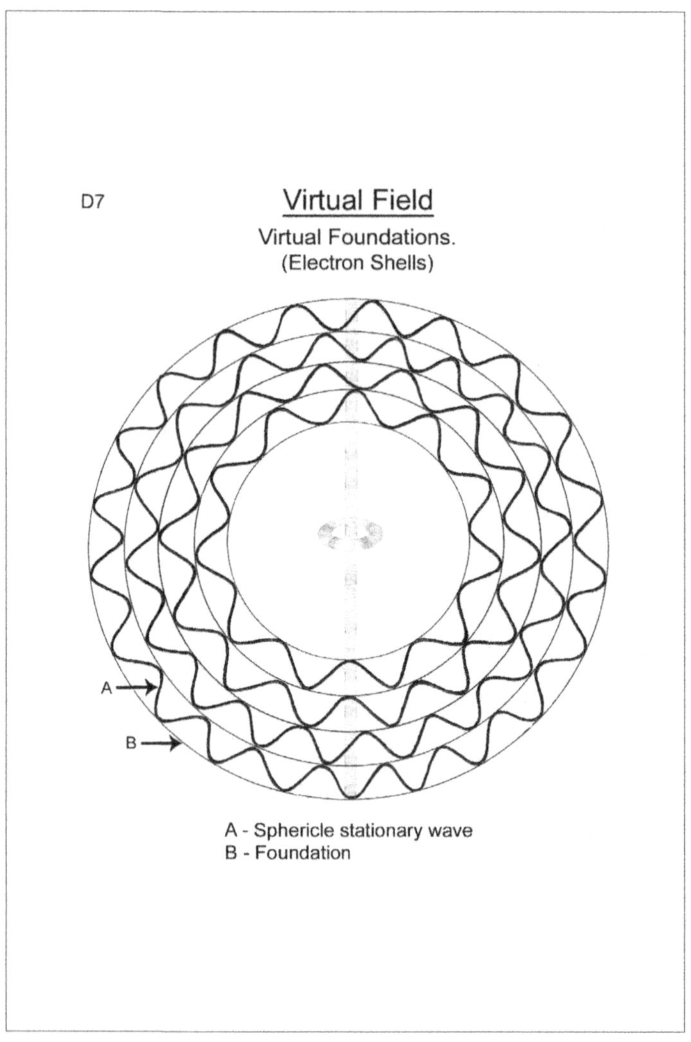

Figure 10

FOUNDATIONS (D7)

The outer atom, which includes the electron cage, is made up of concentric spheres. These Aetheric (onion-like) shells are created by resonant stationary waves, with each layer separated by frequency. The energy for the shells comes from the north/south *impulse current*.

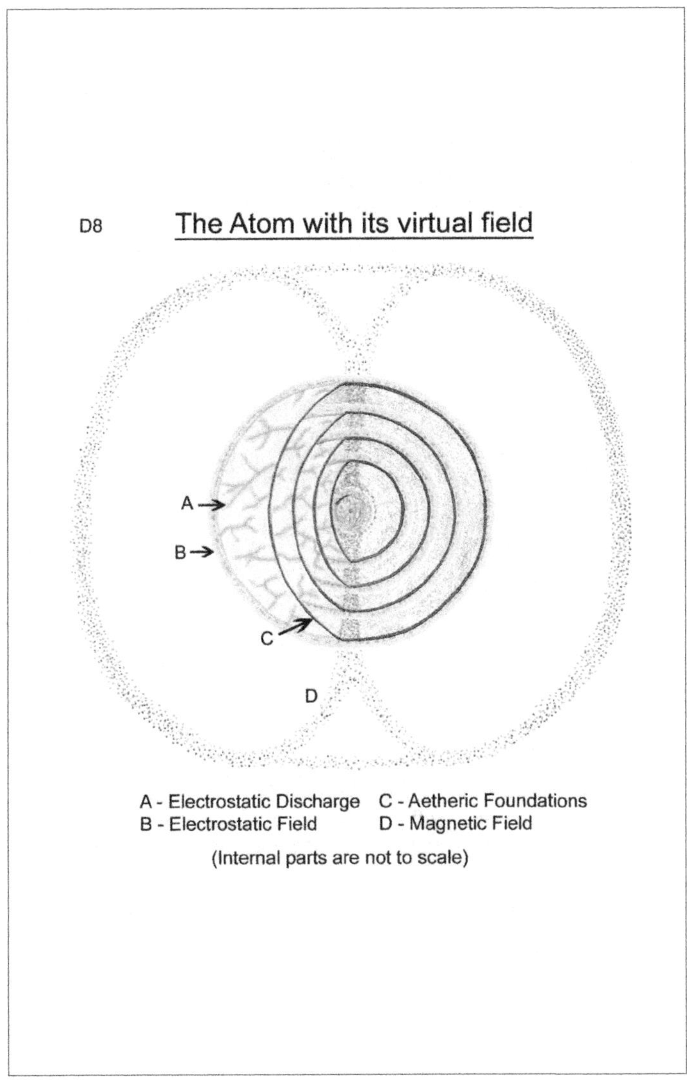

Figure 11

THE ATOM WITH ITS VIRTUAL FIELD (D8)

When observed in its complete form, the atom with its hidden Virtual-Field reveals how important the Aetheric wave structure is. Without the Virtual Field, the atom would be without form or internal structure.

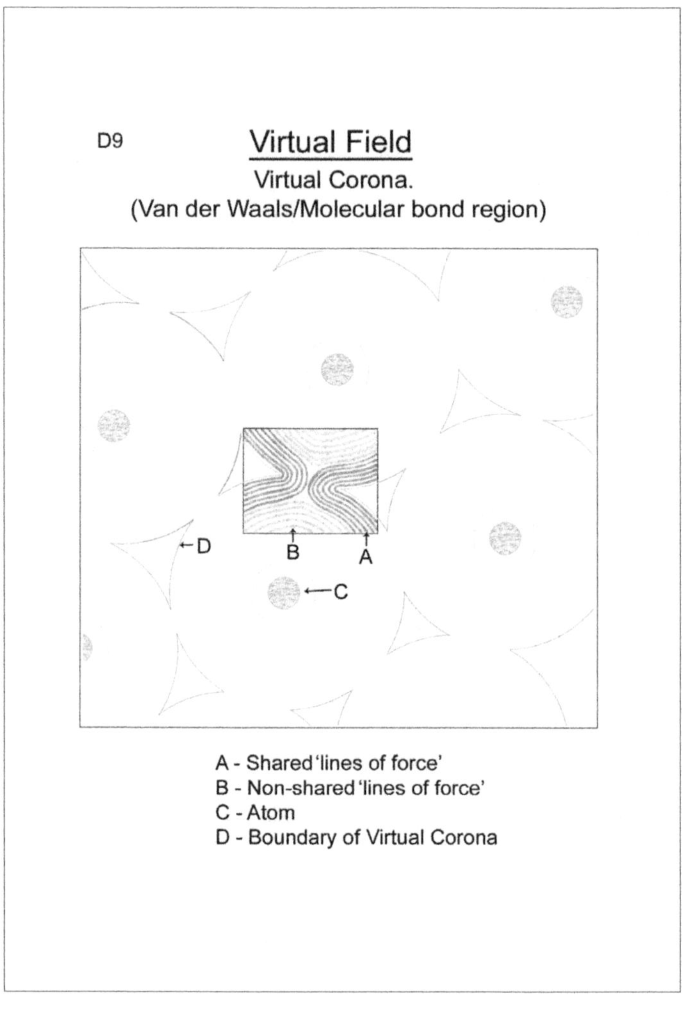

Figure 12

VIRTUAL CORONA (D9)

In Aether theory the Virtual-Corona is the area which surrounds the atom. It serves multiple purposes including interaction between atoms. NOTE: The role the field plays, does not negate other electrical processes, what it does do is describe an underlying structure that helps create real physical objects.

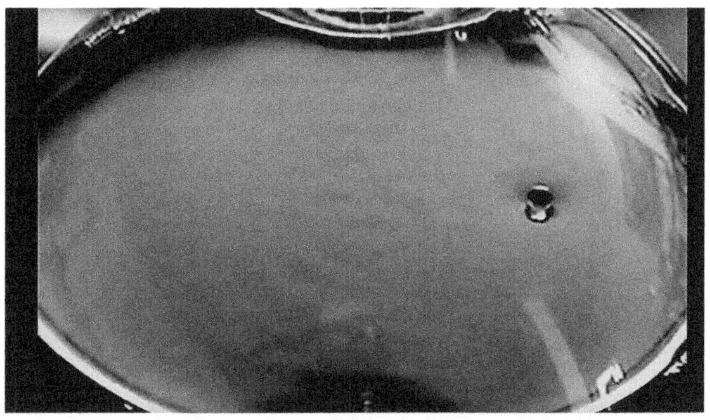

Figure 13 (top): Author and researcher Stan Deyo used nothing more than sound to create spheres in a fluid. By employing a specific frequency of sound, little spheres of fluid appear on top of the main fluidic body. As long as the sound is present, the spheres remain.

Figure 14 (bottom): Here is one example of how sound can produce intricate shapes. The picture of a cymatic wave pattern (left), along with a turtle (right) shows just how complex and diverse sound waves can be.

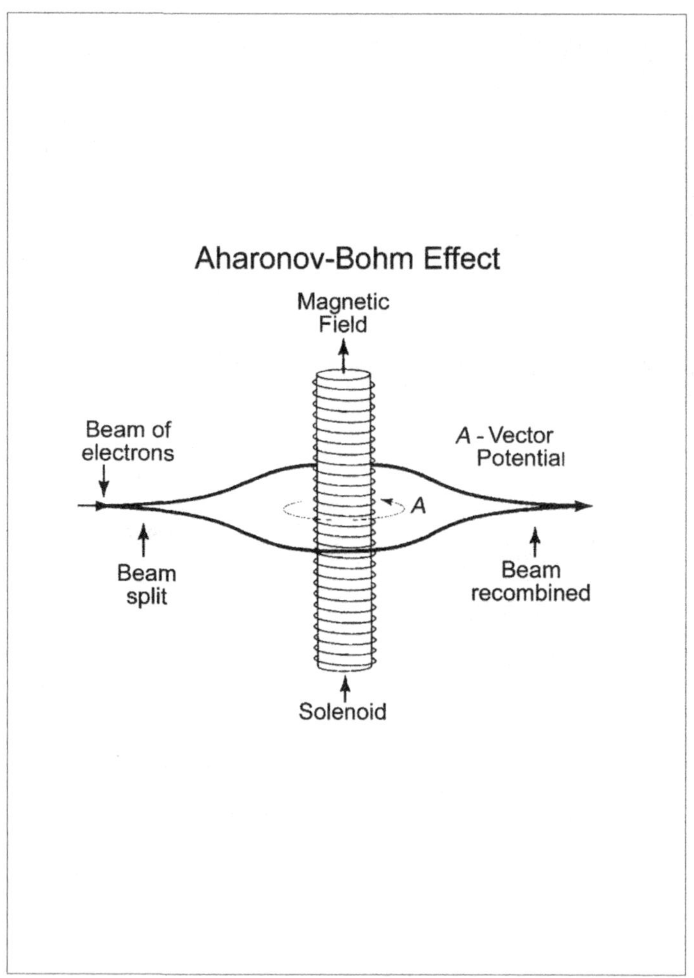

Figure 15

AHARONOV-BOHM EFFECT

In the Aharonov-Bohm experiment, a beam of electrons are sent past a solenoid. They are not affected by the magnetic or electric fields but are instead affected by the so-called non-physical *vector potential*. In this instance, the term vector means it is dynamic (with ordering – when viewed externally), while scalar means static (without ordering – when viewed externally).

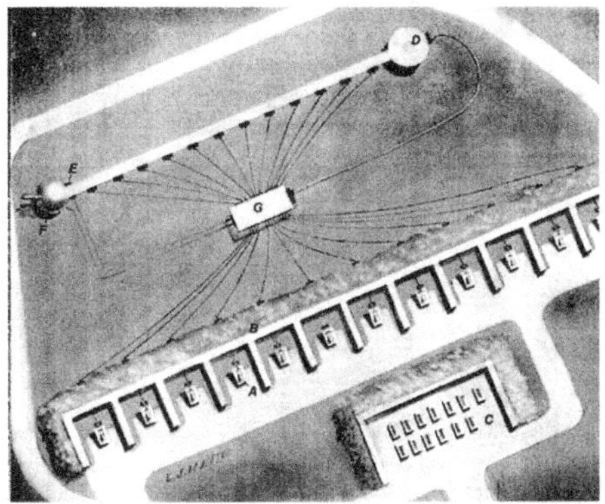

Figure 16: An artist's illustration in *Aviation Week and Space Technology* (July 1980), shows a large weapons facility at Saryshagan missile test range in the Soviet Union (Kazakhstan). Believed by analysts to be some kind of laser, researcher Tom Bearden thinks it might, in fact, be a Tesla Howitzer. Based around Tesla's Deathray concept, Bearden believes it could be used to send multiple scalar waves to any point in or on the globe.

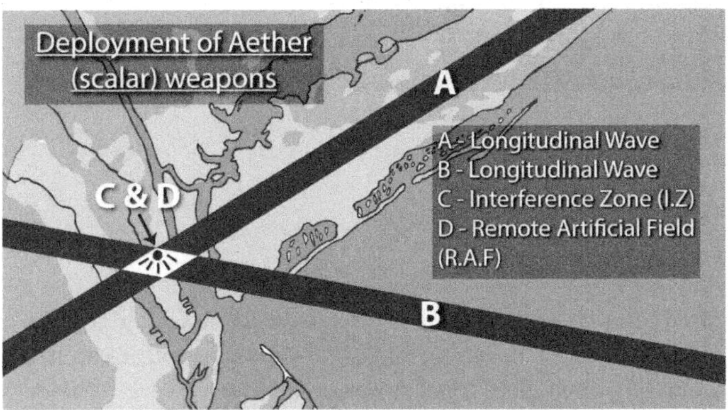

Figure 17:
DEPLOYMENT OF AETHER (SCALAR) WEAPONS
Researcher Tom Bearden describes various ways longitudinal waves are delivered to any point in or on the earth. One way is to send two waves in different directions. The area where the waves meet is called the Interference Zone (I.Z.). Inside the I.Z. is the Remote Artificial Field (R.A.F.), where real electromagnetic effects take place.

Figure 18 (top): Described by the author in the Preface, this formation is known as a "Giant Radial" and appeared almost instantly overhead. The fan like formation dissipated within minutes, and regardless of how it appears in the photo, it was parallel with the ground.

Figure 19 (bottom): Also described by the author in the Preface, the last formation to appear in the sky (between the power pole and left corner tree) was the half-toroid. It was about thirty seconds before it was unrecognisable. These cloud formations appear when the transmitters cut their power to the field, which then allows cold air to rush in and form water vapour. The formations shown in the above photos are called Remote-Artificial Fields (R.A.F.).

Figure 20

CONICAL LIGHTS (Illustration)

In 2007, this peculiar formation was observed by the author (and sister) for twenty minutes and appeared to be quite close to our location. While observing, some of the tubes began to change from an off-white colour to a dark shadow like appearance, and then back again. Could this strange phenomenon be a product of the Aether superweapons?

Figure 21

NIGHT TIME TWIN RADIALS (Illustration)

Seen by the author in 2007, this strange cloud formation appeared overhead at approximately 1.00 am in the morning. With the help of a full moon and the city lights, this enormous cloud moved slowly from north to south and did not lose form during my observation. Researcher Tom Bearden calls these artificial cloud formations, "Giant Twin Radials."

Figure 22: An illustration depicting what photo-journalist Nick Downie saw in 1980 while stationed in Afghanistan. In the foreground are the Hindu Kush Mountains, while the 100 mile wide dome of light rises up behind them.

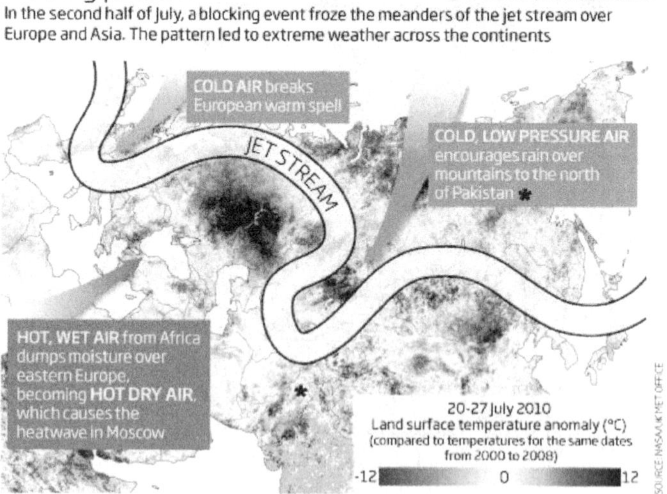

Figure 23: An illustration from *New Scientist* shows the strange twisting of the jet stream over Europe and Asia. By bending the jet streams, normal weather patterns are thrown into chaos over large areas.

CHAPTER TEN

AETHERIC WARFARE

"After this I saw four angels standing at the four corners of the earth, holding back the four winds of the earth, so that no wind would blow on the earth or on the sea or on any tree."

–Book of Revelations, Ch.7:1

WHEN NIKOLA TESLA was developing his T.M.T. transmitter, his vision was clear: he was going to build a system for sending wireless communications, and more importantly power, to any point on the globe. This was so mankind could better itself as a whole, while expanding the independence of the individual. But for others like J.P. Morgan and his kind, the thought of not being able to control and profit from energy and information transfer was not only anti-capitalist, but downright dangerous. It has to be remembered that, to the elite, control and profit go hand in hand, so a system where corporates and government control society together (i.e. fascism) is a good thing. Then there are the military-industrialists like John J. Hammond, who through his commercial relationship with Tesla may have got his hands on the inventor's T.M.T. patents. This second class of secretive and powerful industrialist doesn't just try to predict future events, they try to manage them, and history shows that there has always been a close link between the Morgan's and Hammond's of this world. So yes, while profit is a big part of it, control is the prime directive.

When it comes to Tesla's greatest invention, the T.M.T., it appears it may have initially been suppressed by J.P. Morgan and others out of sheer greed. But its eventual descent into the black world of military intelligence is nothing less than an attempt to create a weapon of frightful potential.

Before describing the physical principles behind the super weapons, it is important to understand how the Aether operates in the natural world, and how it could possibly be engineered for malevolent purposes.

THE NATURAL AETHER

After a short study of the weather, it doesn't take long to see that there are many holes in our understanding of how it works. For example, it's now known by scientists that there is not enough electrical energy in a storm cloud, to produce the lightning discharges we observe. In an article in *Phys.org*, researchers at Langmuir Laboratory (New Mexico Institute of Mining and Technology) state the problems facing current theory.

"It's well known that lightning is an electric current-a quick, powerful burst of charge that flows within a cloud or between a cloud and the ground... This discovery is surprising, since previous simulations have shown that lightning breakdown appears to be negative, meaning the spark moves upward in the cloud from a negative to a positive region. In positive breakdown, the spark moves downward from a positive to a negative region... Currently, the largest electric fields that have been measured inside thunderstorms are several times weaker than what is needed to break down cloudy air and initiate lightning."

Phys.org,
1st March, 2016

Researchers are still mystified at how lightning forms in a thundercloud, the lack of electrical energy (ten times too weak), non-conventional charge flows, and the unknown source for the 'leader' makes current theories untenable.

So, what triggers lightning inside a thundercloud? Mainstream science proposes a number of theories, with one of the leading contenders involving cosmic rays showering down from space. The theory states that cosmic rays bombard the electrical field of a thundercloud, which initiates 'electrical breakdown', and thus produces the 'leader'.

The only problem is, cosmic rays are not powerful enough to trigger 'electrical breakdown', but what if there was another agent capable of initiating the process.

In his Colorado Springs notes, Tesla describes how his giant T.M.T. was designed around the principles of lightning, and that the production of Aetheric-stationary waves was one effect that connected his machine to the powerful arcs. Also, during an acceptance speech in 1917, the inventor details how his wireless power technology in Colorado was able to produce a dense cloud of water vapour throughout his lab.

Could the build-up of Aether energy in a storm cloud's electric field initiate 'electrical breakdown', and thus create the 'leader'? It does appear there could be a link, and it comes in the form of atmospheric ball lightning. While mainstream science is still in the dark concerning its underlying cause, Tesla was quite confident it was Aetheric in origin. If this is so, then it must be assumed the Aether plays an important part in the weather, and Tesla was right to surmise it could be manipulated.

If there is a connection between the Aether and the weather, then there must also be one between the all-pervading field and the ground. When it comes to natural phenomena like earthquakes and volcanoes, the presence of Aether could explain some of the mysteries still perplexing scientists today. For example, why do violent earthquakes occur at depths where the rock isn't solid, but has a consistency similar to Silly Putty? Or why is there a lack of heat inside the 'slip zone' where the rock is under intense frictional pressure? But if that sounds strange, then the following accounts written about by Motoji Ikeya takes it to another level. In his book *Earthquakes and Animals*, Ikeya presents some of those startling accounts.

"The electronic age has added a new category of earthquake precursor reports to ancient legends.

They are accounts of odd behaviour and noises from domestic electric appliances: T.V.s, radios, refrigerators, mobile phones, fluorescent lamps, car navigators, and possibly computers. The reports came in independently from Kobe, Izmut, and Taiwan before the large earthquakes of 1995, and 1999... Interestingly, exposure of some of these appliances to EM pulses in a laboratory reproduced the behaviours described in the reports... Several reports from slightly earlier than the electronic era are also investigated in this chapter: How was it that nails hanging for weeks from a permanent magnet dropped off two hours before the Ansei-Edo (Tokyo) Earthquake (M6.9) in 1855? Why did iron chains in a military factory begin to swing against each other two hours before the Eastern Nankai (M8) Earthquake in 1944? Why did arc discharges occur between iron bars lying on the ground a day or two before the Tangshan Earthquake in 1976? Why could a man no longer read the phosphorescent numbers on his watch 2-3 days before the Great Kanto Earthquake, but see them again one day later?"

–Motoji Ikeya
Earthquakes and Animals,
2004

One interesting thing Ikeya states, is that laboratory tests have proven that electromagnetic pulses can produce the same effects upon electronics as earthquakes do, even if it is only at short distances. But the accounts mentioned lower in the quote point towards another kind of energy, one that's capable of emerging in more subtle ways. Could some kind of Aetheric phenomenon be taking place, and if so, are we able to directly observe this unusual manifestation? The answer appears to be yes, and it comes in the form of Stress Lights.

Known to appear around the time of earthquakes, the mysterious lights usually manifest as 'spheres of light', 'light flashes', or 'misty plasma'. And even though scientists have proposed various theories concerning their origins, their elusiveness makes them extremely hard to investigate.

Interestingly enough, the phenomenon was observed and photographed in New Zealand's capital of Wellington, during the massive 7.8 Kaikoura earthquake of 2016. In a news article at the time, the strange 'light flashes' were seen lighting up the night sky, with scientists only able to guess at what was going on.

> *"Many people overnight reported seeing strange lights in the sky, a phenomenon that has been reported for centuries before, during, and after earthquakes...*
>
> *Much like tidal research, it is an area that is notoriously difficult to investigate. Tidal stresses and their effects on the Earth are minute, but measurable, although many seismologists remain unconvinced by theories of "tidally triggered" earthquakes...*

GNS seismologist Caroline Holden said there were anecdotal reports of lights in the sky. "Unfortunately, we cannot measure this phenomenon or its extent with our instruments to provide a clear explanation," she said...

Yet another theory suggests a link between the electric charge, or current, released by the earth ripping and buckling below the surface, and the magnetic properties of rock.

The charge appears as light, so the theory goes."

**–Stuff News,
14th November, 2016**

Tesla believed the Aether was a mechanical type of energy, so it is able to act upon and move through solid mass in a way electrical energy can't. That's why some of the non-electrical precursor events quoted by Motoji Ikeya point towards the all-pervading field, and not the more commonly held belief of electrical fields. Interestingly, a type of electrical ball lightning has now been produced by scientists in the lab, but their performance and characteristics have no comparison when compared to the eyewitness accounts describing natural ball lightning. The following quote by Gerry Vassilatos gives insight into what Tesla believed about the natural Aether processes in and around the earth.

"Tesla stated that both the geo-internal radio-active elements and the native heat of the earth was a direct result of Aetheric bombardment from space... For Tesla, this revealed a whole

dynamic in which Aetheric flow directed and guided all processes electrical. Aether was the superior, electricity the inferior. What Aether did, electrical phenomena followed. This single principle became for him a rule of unparalleled use. Tesla consistently stated that Aether was being condensed into electrons at a specific rate, by which the terrestrial electrical field was being manufactured."

–Gerry Vassilatos
Secrets of Cold War Technology-
Project HAARP and Beyond,
2000

If the statements made by Tesla are correct, then his experiments at Colorado Springs might hold the key to why phenomena like Stress Lights sometimes appear around earthquakes and volcanoes. While working with his giant T.M.T., the inventor was able to produce powerful resonant effects many miles from his lab. Some of the examples below show the unusual effects taking place.

- Small continuous arcs appeared from lightning rods twelve miles from the lab.
- Incandescent lamps fully lit up many hundreds of feet from the T.M.T.
- Butterflies became entrapped in some kind of an electrostatic whirlwind.
- Various metallic objects produced electrostatic phenomena.

- Sparks would jump from horses shoes to the sand.
- The inventor states that he produced artificial ball lightning at Colorado Springs.

The Serbian mentions more than once that his T.M.T. was designed to mimic naturally occurring Electro-Aether phenomena. What this suggests is, that when events like earthquakes produce intense pressures upon subterranean rock, the build-up of an electric field in the ground creates 'stress', or what is sometimes called a *potential* in the Aether. In turn, the *potential* projects resonance into the surrounding environment, and because it can move through the earth with relative ease, it travels to the surface to produce Stress Lights in the atmosphere.

If the Aether is responsible for the bizarre electrical phenomena seen around earthquakes and storms, then wouldn't that suggest the all-pervading field is an important part of how the earth operates. It might seem strange when first conceived, but for a planet that appears unfathomably complex, there may appear to be a simple mechanism underlying all of its natural processes, and that includes gravity.

If the previous material provides a simple theory of how the Aether and the natural processes of the earth are linked, then Nikola Tesla's T.M.T. was the key to unlocking its incredible potential.

NATURE AS A WEAPON

It is now accepted by most in the mainstream scientific community, that so-called 'empty space' isn't actually empty.

But where the modern active-vacuum differentiates from the Aether, is that scientists don't believe the former could be used to project real power through space. Tesla, on the other hand, stated how he could use his T.M.T. to manipulate the earth's own Aether reservoir, and use the process for sending power and communication signals to any point on the globe.

However, is it really possible to take Tesla's wireless power concept and use it to control the weather, earthquakes and other natural phenomena? As for the inventor himself, he believed that human-induced weather control was possible.

"The time is very near when we shall have the precipitation of the moisture of the atmosphere under complete control."

–Nikola Tesla,
June, 1900

Tesla also declares, it's advancements to his T.M.T. that would enable man to achieve what would appear to be the impossible.

But what about the energies involved, for example, could enough power be harnessed to create an earth-quake? The answer is yes, but it is not achieved by pumping vast quantities of EM energy into the ground, it is done by manipulating the ever-present Aether. The earth's Aetheric reservoir is vast, and the amount of energy Tesla needed to manipulate it, would have been comparable to putting a creek beside the Pacific Ocean. In essence, the Aether energy bound up in and around the earth is for all intents and purposes inexhaustible; all Tesla had to do was open a channel into it through resonance. He even affirmed that the all-pervading field could be harnessed at any point desired:

> *"Ere many generations pass, our machinery will be driven by a power obtainable at any point of the universe... Throughout space there is energy. Is this energy static or kinetic? If static our hopes are in vain; if kinetic—and this we know it is, for certain—then it is a mere question of time when men will succeed in attaching their machinery to the very wheel work of nature."*
>
> *—Nikola Tesla*
> *A.I.E.E.,*
> *1891*

According to the Serbian, not only was the reservoir inexhaustible, but it was engineerable, and if his theories were correct, we would be able to manipulate any part of our natural surroundings.

After his death in 1943, Tesla's wireless power secrets were thought by many to have followed him to the grave. But as it is shown in *Wardenclyffe* (chapter 6), western and eastern intelligence agencies obtained the inventor's secret papers, and may have used them to perfect an incredible weapon.

So, what ultimately happened to Tesla's secrets? According to Lieutenant Colonel Tom Bearden (Retd), a researcher and theorist who has spent decades trying to uncover where Tesla's technology went, believes the Soviets had definitely acquired the inventor's wireless power secrets, and that they have poured vast resources into perfecting them as a weapon.

When it comes to the physics used in the super weapons, Bearden's theoretical work centres around *potentials*, and how they might be more than just mathematical conveniences.

In his book *Star Wars Now! The Bohm-Aharonov Effect, Scalar Interferometry, and Soviet Weaponization*, Bearden explains where he believes the physical principles come from, to make the super weapons possible.

> *"With the advent of Maxwell's equations, electricity and magnetism were combined into an elegant electromagnetic theory, and these equations then served as the basis for the development of modern theory. Gradually potentials were relegated to a position of inferior importance, and they even came to be regarded as purely mathematical conveniences by most scientists.*
>
> *However, with the advent of Aharonov and Bohm's seminal paper, it became crystal clear that potentials are in fact real entities, and they can directly affect and control charged particle systems even in a region where all the fields and hence the forces on the particles have vanished... The Bohm-Aharonov effect shows that the E [electric] and B [magnetic] fields can remain zero, and yet the potentials can still cause physical effects."*

> **–Tom Bearden**
> **Star Wars Now!**
> **1984**

Presented in a seminal paper in 1959 by Yakir Aharonov and David Bohm, the AB effect (Figure 15, page 162) is essentially an experiment where particles are propelled past a solenoid that has zero magnetic and electric field energy radiating out into space. According to classical physics, the particles should pass by the solenoid without a phase shift (a change in the period from one wave peak to the next) taking place. But that isn't what happens, evidence shows that the particles do indeed phase shift, and that the mysterious energy called *potentials* are responsible.

Bearden believes the AB effect proves that the non-physical *potentials* are capable of creating real effects, and that we now have a demonstrable mechanism for engineering the so-called non-physical vacuum (Aether). To help explain the theory of the AB effect and how *potentials* relate to EM fields, imagine the energies involved are analogous to a surfer riding a big wave on the ocean. If the surfer is considered the magnetic (B) field, and his surfboard the electric (E) field, then the *potentials* are the wave the surfer is riding on, and the whole ocean is the all-pervading field (Aether).

According to Bearden, two important papers released over a century ago reveal how the *potentials* could be used to create real electrodynamic effects at a distance.

"Indeed, we assume total primacy of scalar potentials, after the work of Whittaker, holding that all the effects of present electrodynamics can be produced by utilization and interference of two or more scalar potentials."

―Tom Bearden
Star Wars Now!

1984

It's important to note, the AB effect deals with *vector potentials*, while Bearden's work revolves around *scalar potentials*. His description of the 'scalar', in *scalar potential*, simply means a static ordering in the Aether, or to use an analogy, a whirlpool is a static ordering in a rushing river. He also believes that *potentials*, and how they are the foundations of electromagnetism, goes all the way back to James Clark Maxwell. But because his work was simplified, or some might say butchered by Heaviside and Gibbs, the Aetheric component was discarded, leaving modern-day scientists and electrical engineers with an incomplete picture of how electromagnetism works.

Now according to Bearden, two important theoretical papers were released in 1903 and 1904 by Edmund T. Whittaker. They describe how EM force-fields and *scalar potentials* relate to each other, and also how the latter could be engineered to create "spooky action at a distance".

- The 1903 paper describes the existence of a "hidden" set of bidirectional EM waves in the *scalar potential* of the vacuum (Aether). This internal wave structure can be used to change the *potential*, locally or at a distance..
- The 1904 paper describes how all EM fields and waves are composed of two *scalar potential* functions. And that two *scalar potentials* can be turned back into an ordinary electromagnetic field through interference—even at a distance.

In another quote from Bearden, he describes in simple terms how the above process is achieved.

> *"This is somewhat analogous to "putting energy in here" and "extracting it out there" without any travel or losses in between—Nikola Tesla's old 'wireless transmission' of energy at a distance without losses idea... we point out that in theory one may deliberately make a beam containing zero electric and magnetic fields..."*
>
> **—Tom Bearden**
> **Star Wars Now!**
> **1984**

Essentially, what Bearden implies, is that normal EM force–fields can be cancelled at location A, while their *potentials* are sent to another location, and at any speed. Then through interferometry, EM energy can be made to reappear at location B. The scalar energy does not move through three dimensional (3D) space, but travels hyper-dimensionally (4D). To help visualise the process in a simple way, imagine two scalar beams of energy are manufactured and sent out from the transmitter/s, then at a chosen target site, the beams are made to cross. Where ever the two beams cross (called the interference zone), that area is where real EM effects take place. An illustration of this process is shown in Figure 17 (page 163).

Unlike normal electromagnetic waves, which mainstream science is familiar with, Tesla and others made statements about scalar (*impulse radiant currents)* waves that are nothing short of startling.

- With EM waves, the vibrations (oscillations) occur at right angles to the direction of travel. Conversely, scalar waves, produce vibrations (rarefactions and compressions) along the same path as the direction of travel.
- Empty space (the Aether) is packed full of scalar waves.
- Scalar waves do not produce radiation the same way EM waves do, so their movement from point A to point B is totally hidden.
- Scalar waves are capable of speeds ranging from zero to infinity.
- Scalar waves do not fall off at the square of the distance.
- Scalar waves can theoretically move through any 'Faraday cage'.
- Scalar waves can theoretically move information and energy to any distance.

By using advanced versions of the T.M.T., weaponeers could theoretically project real EM energy to any point in or on the globe, and then use that energy to manufacture, steer, or even destroy, hurricanes, tornadoes, lightning, tsunamis, freak-waves, storms, monsoons, fog, frost, cold snaps, permafrost, heatwaves, droughts, humidity, as well as interfere with local air flow, ozone layer, jet streams and even the ionosphere.

Also, they have the ability to manipulate the earth's interior at any point, with artificial effects like tremors, subsidence, ground explosions, and earthquakes, with the latter being used to great effect in warfare. The volcano is another phenomenon that can be used to bring widespread destruction, but even that pales into insignificance when compared to what would happen if a caldera is artificially ignited.

There is much more to Tom Bearden's theoretical work than can be written in this book, and for those who wish to learn more about his interpretation of various scientific papers and journals, a good place to start is with his website at *Cheniere.org*.

Could there be any other pieces of information corroborating the existence of the super weapons? The answer I suspect is yes, because every now and then the wall of silence comes down, and either through over-excitement, or staged propaganda, those in the know make statements pertaining to a weapon of unimaginable power. One of those statements, was made by the Soviet Premier Nikita Khrushchev, on the 15th of January 1960.

> *"Though the weapons we have now are formidable weapons indeed, the weapon we have to-day in the hatching stage is even more perfect and even more formidable. The weapon which is being developed and is as they say in the portfolio of our scientists and designers, is a fantastic weapon."*
>
> **–New York Times,**
> **15th January, 1960**

Khrushchev also mentions that if these new weapons were unrestrainedly used, they would wipe out all life on earth.

The Soviets already possessed nuclear weapons, these were mentioned at the start of the quote, so whatever the weapon was it was not nuclear. Also, Khrushchev was at the heart of the Soviet beast, and he would have known how far his military scientists were pushing back the frontiers of science.

Another Russian politician, who claims his country is in possession of super weapons so powerful that they would leave their enemies totally vulnerable, is Vladimir Zhirinovsky. While he is seen as a hot head by his fellow countrymen, and a 'nut' by western analysts, one of the statements he makes aligns with some of the claims made by Tesla.

"Weapons which nobody knows anything about... not yet. With these weapons, we will destroy any part of the planet within fifteen minutes. No explosion, no ray burst. Not some kind of laser, not lightning. It is a calm and quiet weapon instead, with which whole continents will be put to sleep forever. That's all for now."

–Vladimir Zhirinovsky

While the western media laughs at this man's description of a Russian-owned superweapon, it fits perfectly with how Tesla said his own Aether technology worked. There is no physical disturbance between energy source and energy destination, and while individual nuclear weapons can destroy one city at a time, Aether superweapons can destroy whole continents.

Despite the ignorance shown by mainstream western science, there has been the occasional statement made by officials, concerning the existence of a terrible new kind of weapon. One speculative statement came from the former U.S. Secretary of Defence, William S. Cohen, in April 1997.

"Others are engaging even in an eco-type of terrorism where-by they can alter the climate, set off earthquakes, volcanoes remotely through the use of electromagnetic waves. So, there are plenty of ingenious minds out there that are at work finding ways in which they can wreak terror upon other nations."

—William Cohen

Despite giving the speech during a Q & A session on counter-terrorism, and how it hardly made any headlines, doesn't take away from the fact that he spoke openly about "ingenious minds" devising weapons which can create earthquakes and hurricanes by remote control.

It has been suggested by the website *Debunked* that Cohen's quote was taken out of context, and how it was, in fact, referring to the use of disinformation as a way of tying up resources within western intelligence. But that is not the case, in a 2011 interview with Jim Parisi, William Cohen was asked whether electromagnetic waves were being developed as a weapon to manipulate the weather, in response, Cohen states "absolutely". What he says next is even more telling, he said the weapons are not necessarily the preserve of governments, but could, in fact, be under the control of groups.[1]

When connecting the pieces, it becomes evident that the highly classified technology programs might be controlled by secretive groups outside any government oversight. In the next chapter, *East Meets West*, it describes in greater depth who the groups might be, and how they have absolutely no allegiance to any country, and will use the weapons without compunction.

One final point concerning the veracity of weather control, is how it was brought up by the United Nations for the purpose of outlawing its implementation.

The Environmental Modification Convention, or (ENMOD) as it has become known, was signed on the 18th of May 1977, and ratified by the US in 1979. It begs the question: if these weapons don't exist or are impossible to make, then why would we need an internationally binding agreement to outlaw them. Also, if it was just about simple cloud seeding, which is common knowledge, then why was it based around military activity?

It appears from the previous reports that these weapons may in fact exist, and regardless of whether they are controlled by nations or groups, whoever possesses them holds the balance of military power in their hands.

Interestingly, author and researcher Gerry Vassilatos, in his book *Secrets of Cold War Technology-Project HAARP and Beyond*, believes the fabled HAARP transmitter in Gakona, Alaska, is not a geodynamic weapon. Its design,

---

1 Jim Parisi show, Youtube, https://www.youtube.com/watch?v=NoUmgv7pipo

wave-type and power output makes the facility inadequate for manipulating natural events like hurricanes or earthquakes. He also claims the Eastlund patents are a ruse, and that they have nothing to do with Tesla's *impulse current* (scalar wave) wireless power system.

When it comes to the Aether superweapons, the power of the biodynamic and psychodynamic aspects cannot be underestimated, but it is the geodynamic portion that gives us the greatest evidence of their existence. Tom Bearden believes the superweapons have been used many times over the last fifty years, and that mainstream science has been oblivious of their surreptitious use.

While overlooking certain aspects of the technology, like the remote detonation of fuel and powder, or the distant killing of electronics and melting of metals. It is the following five examples that provide the best indication of what the weapons are capable of, with the first of these occurring around 1978, in what has become known as the Bell Island Boom.

BELL ISLAND

This account is based on a documentary from
CBC NEWS (1978).

On the second of December 1977, a series of booms began occuring along the North American coast, with their locations stretching all the way from Georgia to Nova Scotia in Canada. These booms, or what some have dubbed 'air quakes' rattled citizens so badly, the government had to call in the military to try and find an explanation. After investigating various theories, the booms were put down to unusual atmospheric conditions carrying sonic booms from distant jet fighters. This information was later shown to be pure conjecture.

Three months later, on the 3rd of March 1978, the booms stopped for no apparent reason, then resumed again later that month. Interestingly, one expert from Lamont Doherty Observatory noticed how the booms seemed to take weekends and holidays off, as if they were on some kind of timetable.

Then on the 2nd of April 1978, a strange set of events took place on Bell Island, Newfoundland that made the previous four months seem normal. On a hazy Sunday morning at around 11 am, a huge boom occurred that was heard 37 miles (60 kms) away, but it's what happens next that leaves the locals dumbfounded, and authorities scrambling for answers. It appears that a massive discharge of energy occurred on the island, and from the bizarre eyewitness accounts, it could only have been Aetheric in nature. Below it describes some of those events.

- Flames leapt out of power sockets;
- Lights came on by themselves;
- The hands of an old wind up clock began to move around quickly, which was a small miracle considering it hadn't been working for years;
- Television sets smoked and exploded in different areas;
- Small ball lightning came out through oven windows;
- Three-foot wide ball lightning was observed up close;

- Three different buildings were hit by something, leaving electronics destroyed and buildings physically shattered, but not burnt;
- Chickens in a coop were killed, they had blood running out of their eyes and beaks, but didn't suffer from burns;
- Three mysterious holes appeared in the ground at different locations;
- Many people claimed there was a strange tone heard before the giant boom took place;
- Some claim to have seen a strange light hit the island from above.

Just as Canadian authorities turned up to investigate what appeared to be a meteor strike, they discover that foreign investigators had come in from New York state and Los Alamos National Laboratory in New Mexico. This was odd, because the Americans were asking the local Canadian investigators if they had security clearance to be there! When asked how they got there so quickly, the Americans said they already knew, but when pressed further, they said their satellites were watching. Later it was discovered by Canadian authorities that the Americans had not been cleared to enter Bell Island.

If that wasn't strange enough, the whole episode took a turn for the surreal, when an official statement came back from the authorities concerning the event. What they said was, a high altitude thunder storm had occurred that people on the ground were unaware of, and from that storm came a giant lightning bolt. The electrical discharge was one of the biggest ever recorded, and was responsible for all the phenomena observed. Continuing on, they also say how there may have been some kind of Soviet mind weapon involved, causing mass hallucinations.

The official explanation is absurd, and for the American and Canadian governments to blatantly lie about what happened on the east coast of North America over a four-month period is scandalous. It does raise the question though, were the elected officials of both countries even aware of what was going on? Probably not.

So, what did happen along the east coast and in particular Bell Island? The only feasible answer is, they were deploying a more advanced version of Nikola Tesla's T.M.T. developed at Colorado Springs. In my opinion, they were fine-tuning a giant Aetheric-stationary wave, that is still in operation today, which allows them to create artificial weather modifications. When we look at the events on Bell Island, the observed evidence points to an accidental discharge of Aether energy.

The resulting effects – like the illumination of light bulbs, and the destruction of appliances – are the result of Aether-*potentials* downloading into the electrical circuits and creating powerful voltage spikes. Also, the mechanical clock moves because the Aether is a mechanical force that can act upon non-electrical systems. As for the ball lightning, it is an Aetheric phenomenon that Tesla produced with relative ease. Lastly, because there was no autopsy, the incident with the chickens remains a mystery.

If one wishes to use EM energy to explain the unusual events at Bell Island, it will forever remain unsolved. The only way to decipher the mystery is to apply Aether physics, and its weaponization after World War Two.

In more recent times, people have heard and even recorded strange low bass rumbling sounds that sometimes go on for hours. These I believe, can be caused by giant stationary waves when they are being adjusted, which results in the build-up of resonance in the ground and atmosphere. When it comes to the people who hear these haunting sounds, the feeling they get leaves them extremely unsettled to say the least.

THE NORTH PACIFIC EXPLOSION

In 1984, reports of a strange event in the Far East began to circulate in the west. The story was eventually picked up by the *Chicago Tribune*, who then wrote a piece about a massive explosion near the coast of Japan. In the article, Toby Grotz, the president of the *International Tesla Society*, gave his opinion as to why the explosion displayed many unusual characteristics.

> *"Of particular interest to Tesla researchers, said Grotz, is a widely reported April 9 1984, event in which at least four airline pilots reported seeing an eruption near Japan that appeared to be a nuclear explosion cloud that billowed to a height of 60.000 feet and a width of 200 miles within just two minutes and enveloped their aircraft.*
>
> *In late July, the Cox News Service reported that all four planes had been examined by the U.S. Air Force at Anchorage, Alaska, and were found to be free of radiation despite the fact they had flown through the mysterious cloud in question.*
>
> *Grotz said that such clouds could form if*

someone were attempting to implement Tesla's plans for broadcasting energy by "creating resonances inside the earth's ionospheric cavity" calculated in Colorado Springs during 1899 experiments by the electrical genius."

**–Chicago Tribune,
10th August, 1986**

One thing is certain—it could not have been a conventional nuclear explosion or the passenger jets would have dropped from the sky because of the electromagnetic pulse. So what was it? According to the article, which includes a paper from Tom Bearden, the event was a weapons test by the Soviets. Bearden states, the scalar (*impulse current*) weapons are not only capable of dumping energy into a given area, causing a hot explosion (Exothermic mode), but they can also pull energy out, triggering a cold explosion (Endothermic mode). The cold explosion is initiated when all the EM energy is pulled out of the 'interference zone', causing water to come up from the ocean, and violently mix together. It is at this point that the giant cloud of super-cold vapour expands outward and upward. Bearden goes on to say how the non-existent flash of light, lack of radiation and shock-wave, implies the 200-mile wide mushroom cloud was a cold explosion, and if an invading army was caught inside the 'interference zone' (the jet planes were outside the I.Z.) they would be instantly turned into ice.

A weapon that can deliver such power without the need for a physical bomb, is truly revolutionary, and because there is no way of shielding against it, an enemy would be totally vulnerable to an attack.

It's interesting to note, that, in more recent times, airline pilots have reported strange light shows in an area just north of this event. Known as the Kuril Islands/Kamchatka region, it is tightly monitored by the Russians, making one wonder if it isn't a test range for exotic weapons.

THE HINDU KUSH LIGHTS

On the 17th of August 1980, an article (non-related illustration on page 166) appeared in the London *Sunday Times*, describing what war cameraman Nick Downie observed while stationed in Afghanistan. In the article, he states how he saw giant domes of light manifest from deep within the Soviet Union, and that they were so big they could be seen from the opposite side of the Hindu Kush Mountains. He estimated their size to be well over one hundred miles wide, and as they grew in size their light became less intense.

What did Nick Downie see deep within the Soviet Union? According to those who have investigated this event, it was a weapons test of some kind by the Soviets. And the only credible option, is that it was based on Tesla's wireless power theories. If that is the case, then the spheres are industrial sized versions of ball lightning, and are created by locking together two or more scalar waves, which in turn creates an EM globe of light.

Depending on what the operator requires, globes can be customized for a myriad of tasks. For example, they can be formed into a hemisphere and placed over a sensitive target like a city or weapons site, and if the energy is strong enough, incoming objects will literally vaporize. This would make it impossible for incoming planes, tanks, missiles, and even nuclear explosions to penetrate, making it the ultimate ABM (Anti-Ballistic-Missile) shield. Also, the domes/globes can be placed at any point around the planet for protection against enemy weapons.

Over the last forty years, these domes/globes have also been reported in various locations such as Australia, the UK and Sweden.

THE 2010 WEATHER SPIKE

The world's weather is like a naturally flowing river, controlled by the sun, the earth, and most definitely the Aether field. By manipulating the Aether on a local scale, the superweapons can be used to control individual events like storms, tornados and even hurricanes.

But those events pale into insignificance when compared to what the weapons can do on a continental or even hemispherical scale. Their ability to turn the weather into an unpredictable chaos may have occurred when a very unusual blocking event took place across parts of the northern hemisphere. The weather anomaly spanned Central Europe, China, Japan and India causing unprecedented upheaval, but it was events inside Russia and Pakistan that caused the most destruction. Taken from an August 2010 article in *New Scientist* (and shown in Figure 23, page 166), it describes what took place before and during a major jet-stream diversion.

"Raging wildfires in Western Russia have reportedly doubled average daily death rates in Moscow.

Diluvial rains over Northern Pakistan are surging south—the U.N. reports that 6 million have been affected by the resulting floods.

It now seems that these two apparently disconnected events have a common cause. They are linked to the heatwave that killed more than 60 in Japan, and the end of the warm spell in Western Europe. The unusual weather in the U.S. and Canada last month also has a similar cause.

According to meteorologists monitoring the atmosphere above the northern hemisphere, unusual holding patterns in the jet stream are to blame. As a result, weather systems sat still. Temperatures rocketed and rainfall reached extremes. Renowned for its influence on European and Asian weather, the jet stream flows between 7 and 12 kilometres above ground ...

In recent weeks, meteorologists have noticed a change in the jet stream's normal pattern.

Its waves normally shift east, dragging weather systems along with it. But in mid-July they ground to a halt... There was a similar pattern over the U.S. in late June ...

And so it was that Pakistan fell victim to torrents of rain. The blocking event coincided with the summer monsoon, bringing down additional rain on the mountains to the north of the country. It was the final straw for the

Indus's congested river bed. Similarly, as the static jet stream snaked north over Russia, it pulled in a constant stream of hot air from Africa. The resulting heatwave is responsible for extensive drought and nearly 800 wildfires at the last count. The same effect is probably responsible for the heatwave in Japan, which killed 60 people in late July. At the same time, the blocking event put an end to unusually warm weather in Western Europe.

Blocking events are not the preserve of Europe and Asia. Back in June a similar pattern developed over the U.S., allowing a high-pressure system to sit over the eastern sea-board and push up the mercury. Meanwhile, the mid-west was bombarded by air from the north, with chilly effects. Instead of moving on in a matter of days, "the pattern persisted for more than a week" says Deke Arndt of the U.S. National Climatic Centre in North Carolina."

**–New Scientist,
10th August, 2010**

Officially, this extreme 'blocking effect' was put down to the sun's influence on the Rossby waves (counter-currents to the jet streams), but the severe nature of the effect over such a short period defies all logic. In my view, only the scalar superweapons and their giant stationary waves could adequately explain the interference imparted upon the jet stream winds across the northern hemisphere.

It's interesting to note that the 'blocking effects' also appeared earlier that year over North America and Western Europe, causing anomalous weather phenomena, like unseasonal heatwaves and particularly severe storms.

While circumstantial evidence points towards the use of sophisticated weather weapons, not a lot has been forwarded concerning their use inside the ocean. It does beg the question however: what happened in early 2010 to cause the giant Atlantic Gulf Stream to abruptly change its course from its normal Caribbean/Norway route to a westerly Caribbean/Greenland direction? Was it just coincidence, or was it under the same control as the jet-streams? Whatever it was, it left the scientific community collectively scratching their heads.

Through the use of powerful Aether superweapons, we now have the ability to engineer the world's weather by interfering with the jet-streams at specific locations. It doesn't leave much to the imagination when we consider how effective this form of warfare would be against an unwitting populace, regardless of how powerful they think their armies are.

The following points describe what happened to some of the nations involved, and how those events can affect national, as well as global food supplies.

In Brief

- Russia declares a state of emergency as raging wildfires (nearly 400 within 24 hours) devastate their already drought-stricken crop regions.
- Russia orders an immediate halt to selling off

their grain reserves, because of a 26% drop in production due to drought.

- Russia's worst heatwave, along with devastating floods in China and India cause global food prices to skyrocket.
- Catastrophic floods hit China killing nearly 1,000 people, leaving nearly 600 missing and causing tens of billions of dollars worth of property damage, even threatening the world's biggest dam (The Three Gorges Project).
- China had to cut its forecasts for rice and other food production due to the worst flooding in years.
- Catastrophic flooding on the sub-continent, killed 40 in India's richest grain region.
- Pakistan received so much rain that 20% of its land area was affected by flooding, and with nearly 3,000 people dead, the country is considered a disaster zone.

CHRISTCHURCH EARTHQUAKE SWARM

Starting in the early hours of September the 4th 2010, a major quake struck the city of Christchurch, the second largest metropolitan area in New Zealand. When it was over, four major quakes and 10,000 aftershocks had rocked the city for fifteen months, and estimates suggest, it will take at least a decade to fix the ruined infrastructure. To the civilian populace this was devastating, but to the scientific community, the quakes caused a different type of problem. Many have been left baffled by how the earthquakes struck and also by the way the energy was distributed throughout the city and province.

While New Zealand sits on the Pacific Ring of Fire, the city of Christchurch is located about 100 kilometres (62 miles) from the main Alpine fault. Previous quakes have damaged parts of Christchurch before, but they are believed to have originated far from the city. The largest of these rocked the distant mountains in 1888, but because they lacked accurate sensing equipment, it's hard to know what long-lasting effects it left. Fast forward 122 years and the situation is quite different; with modern equipment, we can now obtain an accurate account of what is going on deep underground.

While the earthquake of 1888 was certainly natural, the quake swarm of 2010/11 definitely showed signs of some type of interference. The initial piece of evidence came when the first major quake hit on the 4th of September 2010. With its epicentre 20 kms (13 miles) west of the city, why did a disproportionate amount of damage occur in the city's eastern suburbs? Authorities said, it was because certain rock types in that area built up extra energy compared to the rock around it. Later, Canterbury University delivered a press release, stating how they couldn't understand what happened when the Greendale (official epicentre) fault slipped. This led scientists to make some wild theories about why that particular quake sent energy outwards in such an unusual way.

"Scientists are investigating whether the 7.1 magnitude earthquake which struck Christchurch and its surrounding communities today was actually two or three shocks in quick succession... scientists are still trying to

reconstruct the way today's quake played out. "We think that this is a very complex event" said geophysicist Paul Caruso. "We think that the main shock may have consisted actually of three earthquakes."

G.N.S. science in Wellington said it could not confirm the mechanism of the earthquake. "There are several parts to this earthquake occurring within seconds of each other and it will take some time to decipher what the wave forms recorded by our seismographs tell us about the sequence of events" said a spokesman.

Professor Kevin Furlong said there was an initial foreshock of about a 5.8 magnitude, about five seconds before the main impact, and possibly from a slightly different location. "I think a lot of people were woken up by that, without knowing why, then whammo.

"The main shock itself had two main pulses of energy in it."

**—Stuff News Website,
4th September, 2010**

In December, scientists went on to state how ten times more energy was released across the province than was produced by the slipping Greendale fault on its own. The only problem with this discovery was, there were no other quake epicentres registered. If the official explanation is to be believed, then nothing about the September earthquake makes sense. The only rational explanation for the excess energy and unusual damage pattern was an artificial field of Aether placed under nearby Pegasus Bay. It built up pre-stressed energy in the ground before the actual earthquake at Greendale took place.

The second big earthquake hit on the 22nd of February 2011, at about 1 pm in the afternoon, and left a death toll of 185. That quake also showed signs of manmade manipulation. Firstly there was a strong tremor practically on the surface earlier that morning, then when the big one hit it was only 8 kms (4.9 miles) from the CBD (Central Business District) and 5 kms (3.1 miles) below the surface.

Another unusual effect that surprised scientists was the way the earthquake dispersed its energy. Instead of the normal 360° fan-like distribution, this quake sent most of its energy in one direction, just like a shot gun, and that direction just happened to be towards the CBD, coincidence? Not likely. The shot gun effect created one of the highest gravity potentials (up/down and sideways motion) ever recorded, and if it had continued on for another 10 seconds it would have left the city flat. All this from a quake that only registered 6.3 on the Richter scale.

The final big quake struck on the 23rd of December 2011, and was situated just off the coast in Pegasus Bay. It too showed signs of an irregular nature not common with earthquake behaviour. Scientists were surprised when they found that the aftershocks that followed the quake had their epicentres located in a way that resembled an explosion rather than an earthquake fault. Stranger still, they admitted to changing the positions of the epicentres, why would they do that? It was as if the Aether energy held in the ground was being released for the last time.

A final note concerns several eyewitnesses who observed

unusual light phenomena. Some say they saw huge flashes of light brighten up the pre-dawn sky on September 2010, while others had reported a purplish aurora type glow before the same event.

According to researchers like Tom Bearden, there is nothing new about man-made earthquakes, but what makes this program so different, is the sustained assault that took place in one location over fifteen months. Because of the four major events and 10,000 aftershocks, it was like the city was under a constant barrage of gun fire. Then there is the non-physical, psychological damage, that has affected many over the ensuing years, with people suffering from severe anxiety and deep depression.

So, what can be accomplished by orchestrating an artificial natural disaster? To start with, there is the financial burden placed on any city or nation in need of new infrastructure, and that the affected area will require loans from lending institutions, foreign countries, reinsurance, or government-issued bonds. Either way, it is the people who will carry the burden of financing the rebuild. An example can be given with the Christchurch quakes, they will end up costing the city and country $40 billion, and that's from a population of about 5 million people.

Secondly, the political and social cost these types of disasters create can leave a nation vulnerable. That's because they lead to instability in the political system, especially when there are big changes in population demographics, like what was seen in the Haitian earthquake of 2010. A city or nation that receives a large enough event, will open the people up to any request the government, or even an outside agency like the UN, might impose, even if they suggest martial law by foreign troops.

In military terms, the technology is more powerful and stealthy than anything in man's inventory, giving its owners the ability to change the whole world politically and economically. And while it's doubtful these weapons would be given sole charge of taking out the established political systems in exchange for a 'New World Socialist Order', it is believable to suggest they are a powerful tool in helping to achieve that goal.

So, who is in control of these weapons? That is a road which leads to a destination most people couldn't predict. That's because a lot of what we have been taught through the media is either half-truths, or downright lies, and even though the truth is out there, it can only be found when a person is prepared to forgo some of the long-held beliefs they hold so dear.

CHAPTER ELEVEN

EAST MEETS WEST

"As long as the Soviet Junta will keep on receiving credits, money, technology, grain deals and political recognition from all these traitors of democracy or freedom, [then] there is no hope for changes in my country [the Soviet Union]. And the system will not collapse by itself, simply because its being nourished by so-called American imperialism. This is the greatest paradox in [the] history of mankind, when [the] capitalist world supports and actively nourishes its own destroyer."

–Yuri Bezmenov, Soviet Defector,
KGB Propaganda Agent, 1984

AFTER TESLA'S DEATH in 1943, United States intelligence was able to examine the inventor's technical papers. Then by the end of the 1940's, the Soviets had also obtained certain portions of his work. If this is true, then it raises an interesting question: if highly classified technical information made its way into both US and Soviet intelligence, and they are truly enemies, then why hasn't one nation destroyed the other? The answer is simple, yet it is one that is quite hard for people to accept. To identify those who possess the superweapons, it must first be understood that both the eastern and western intelligence communities may, in fact, be controlled by the same people.

Regardless of what we've been taught by our educational institutions and mass media, there are only two dominant intelligence systems operating in the world today. The first is the western system, which was created through international agreements, clandestine operations, and the influence of Trans-National Corporations. An example of this unity can be seen with the ECHELON system, a co-operative intelligence gathering arrangement between the US, UK, Canada, Australia and New Zealand. Western intelligence might have many elements, but for the most part, it is dominated by a core of powerful US agencies.

What about Russia and China? As we will see further in the chapter, eastern intelligence originated in Soviet-Russia, and its dominance over nations like China, Cuba and North Korea is still evident today.

But are the two intelligence systems linked? There does appear to be one area where a connection is exposed, and that is through the hidden hand of the *financial elite*. That financial power was there when Russia's Royal House of Romanov was facing up against the communists in 1917. Historians tell us that the establishment of communism in Russia was the result of localised rebellion against draconian rule by the Romanovs. But that is only part of the story. The final events surrounding the fall of the royal family are far more sinister, and they were not brought about by internal conflict.

SOVIET ERA

The official version of events tells us that a group of revolutionaries called the Bolsheviks rode a massive wave of popular discontent that was consuming the country at the time. Then as other factions were pushed aside, the Bolsheviks were able to take control of Russia by November 1917. While that account is essentially true, there are other less well-known details that bring suspicion to the legitimacy of a home-grown revolution. They include:

- By March 1917 Czar Nicholas II had already abdicated from the throne, and his power had been handed to an interim government led by Alexander Kerensky. If the Kerensky government had established itself permanently, Russia would have become a western-style democracy.
- Between March and November 1917, the fledgling democracy had turned into a communist dictatorship, despite the direction it had been moving in.
- A few months before the November revolution, Vladimir Lenin was still exiled in Switzerland for attempting to overthrow Russia in 1905.
- The communist agitator Leon Trotsky wasn't in exile, but he was living outside of Russia a few months before the revolution. Before entering Russia, he worked for a communist paper in the United States and was then arrested by Canadian authorities when he travelled to Nova Scotia. The story goes that after Canadian authorities had

him imprisoned, the British got him released, and from there he went to Russia.

There is an old saying, "A revolution never happens in a vacuum". So, who had the political and financial power to overthrow the Kerensky government and the direction Russia was moving in? In his book *None Dare Call It Conspiracy*, Gary Allen uses a quote from White Russian General Arsene de Goulevitch, the founder of Union of Oppressed Peoples in France, to confirm who was behind the communist revolution in Russia.

> *"The main purveyors of funds for the revolution, however, were neither the crackpot Russian millionaires nor the armed bandits of Lenin. The 'real' money primarily came from certain British and American circles which for a long time past had lent their support to the Russian revolutionary cause... The important part played by the wealthy American banker, Jacob Schiff, in the events in 'Russia' though as yet only partially revealed, is no longer a secret."*

> **—Arsene de Goulevitch**

It is believed that Jacob Schiff, sent twenty million dollars worth of gold into Russia to help the Bolsheviks overthrow the Kerensky government. Other famous banking names like Warburg and Rockefeller also sent millions of dollars to support the bloody revolution. Had it not been for outside help from groups like the international bankers, the Bolsheviks could never have formed a revolutionary army, let alone take over a vast country like Russia. It is a complete lie that communism is a movement of the masses against the Bourgeois. To the contrary, it is an extremely expensive undertaking, that requires smart coordination of many assets.

Another important aspect of the 1917 revolution was how the communists became financially destitute with very little food or fuel. This made it almost impossible for them to extend their influence out from St Petersburg. It is believed by some researchers that the Bolshevik revolution would have died off quickly without outside help. That help came from two sources, with the first involving aid from the US government.

Created in 1919, and headed by Herbert Hoover, the American Relief Mission was tasked with providing aid to war-torn Europe. Then by 1921, it had grown to include Russia, particularly areas devastated by civil war and famine. Now, according to Dr Stanley Monteith's book, *Brotherhood of Darkness*, Hoover kept a catalogue of the amounts given to both the anti-communist and pro-communist areas within the country. Below is that list.

Food, clothing, and medical supplies:

- 27,588 tons sent to the areas controlled by the anti-communists.
- 740,571 tons sent to the areas controlled by the Bolsheviks.

Charity from the United States:

- $332,508 sent to the areas controlled by the anti-communists.
- $55,994,588 sent to the areas controlled by the Bolsheviks.

The level of aid provided to the pro-communist Bolsheviks was vastly disproportionate to the amount given to the anti-communist forces. This leaves us with only one obvious conclusion, and that is, the pro-democratic forces were never going to be allowed to win, no matter how much internal support they received.

The second source of help came from the international bankers, through men like Max May. Who was Max May? He was the vice president in charge of foreign operations for Guaranty Trust of New York, the largest trust company in the United States, and controlled by J.P. Morgan Bank. While working at the Hoover Institute (Stanford University), Professor Antony C. Sutton wrote a series of books based on classified files he had been given short-term access too. In one of those books, called *Wall Street and the Bolshevik Revolution*, Sutton details the financial relationship the communists had with the international bankers after the 1917 revolution.

"In early October 1922, Olof Aschberg met in Berlin with Emil Wittenberg, director of the National bank fur Deutschland, and Scheinmann, head of the Russian state bank. After discussions concerning German involvement in the Ruskombank, the three bank-

ers went to Stockholm and there met with Max May, vice president of the Guaranty Trust Company. Max May was then desig-nated director of the Foreign Division of the Ruskombank... Guaranty Trust used Olof Aschberg, the Bolshevik banker, as its interme-diary in Russia before and after the revolution. Guaranty was a backer of Ludwig Martens and his Soviet Bureau, the first Soviet representa-tives in the United States. And in mid-1920, Guaranty was the Soviet fiscal agent in the US; the first shipments of Soviet gold to the United States also traced back to Guaranty Trust."

–Antony C. Sutton
Wall Street and the Bolshevik Revolution

The Soviets were bankrolled from the start through loans from the international bankers and their agents in the US government. Companies like J.P. Morgan were critical in funnelling resources back and forth between the West and the Soviet Union. This arrangement was essential in sustaining Russian communism throughout its seventy-year history.

The *financial elite* have gone to great lengths to make sure eastern communism survived, and as long as they kept control of a small clique within the Soviet government, they would remain the hidden hand behind global communism.

KGB

All communist (and fascist) systems are authoritarian in nature and are ruled by a very small clique at the top. It is assumed by most people that the Politburo controlled the Soviet Union during the communist era and the Duma has ruled since its fall in 1989. But according to one Soviet defector, the KGB (Cheka-KGB-FSB) has ruled Russia since the time of the communist revolution, and had no intention of relinquishing power after the end of the Soviet era.

In 1961, a KGB Major, Anatoliy Golitsyn, handed himself and his family over to the CIA while stationed in Helsinki, Finland. He was one of the highest ranking officials to ever defect from the Soviet Union. The Major was responsible for the direct and indirect capture of many Soviet spies in the West, including the notorious Kim Philby. But what made Golitsyn stand out from other defectors, was that he was a member of the KGB's Strategic Planning Department, which oversaw future strategy.

Golitsyn told the Americans that he was an 'insider' within the KGB and that his particular department was charged with devising ways of covertly destroying the West.

From amongst his statements, two important revelations were given on how the Soviet Union would execute their plan. The first was that the China-Soviet split (1960–1989) was nothing more than a ruse to deceive the West into believing eastern communism was divided. This is important because it shows how there is no real divide within eastern intelligence. And secondly, one day the Soviet Union would declare itself 'defeated', and thus introduce democratic reforms (Perestroika). This too was a ruse. It's at this point that the CIA began to distance itself from Golitsyn, and that's the reason why he wrote his two books about the communist 'grand plan' (*New Lies for Old*: 1984, *Perestroika Deception*: 1995).

The fall of Soviet communism in 1989 was supposed to bring humanity into a new era of peace and prosperity, but according to Anatoliy Golitsyn, the ruse was to make the Soviet regime appear more democratic while keeping the true power within the KGB. He states how many KGB officials were going to 'supposedly' give up their communist ideologies, and take up leading positions in the newly created democratic states of eastern and central Europe. That statement has now been born out as fact in a recent article by *Russia Beyond the Headlines*.

> *"The Committee for State Security (KGB) was abolished 25 years ago, but its former members still administer Russia. They are deputies, senators and heads of state corporations, whose "Chekist" past has only helped their careers. On the other hand, all of Russia's security organs are the heirs of the KGB."*

> **–Russia Beyond The Headlines,**
> **28th June, 2016**

Anatoliy Golitsyn's earth-shattering revelations about a secret Soviet plan, are critically important for two main reasons. Firstly, we in the West have been duped into believing Soviet communism is dead, and also how the left-wing socialist agenda sweeping the western world today is not Soviet in origin.

Secondly, the true core of power throughout the Soviet and post-Soviet era was and is the KGB (Cheka-KGB-FSB), and its plan is nothing less than the complete destruction of the western world. Golitsyn called himself a KGB 'insider', confirming the fact that there is an outer-ring of agents and officials, as well as an inner-core that controls everything, including the government. As it was pointed out, western bankers helped finance Soviet communism from the beginning, and they still rule it through the KGB. But where does China fit in, is that vast nation truly independent, or is it nothing more than an agent for the *financial elite*?

CHINA

If Anatoliy Golitsyn was correct about the close alliance between the Soviets and Chinese, then it gives credence to the argument that a unified communist front is out to destroy the West. But what about the *financial elite*, is there any proof linking them with the establishment of Chinese communism? The answer appears to be yes, and it comes in the form of a 1952 US Senate investigation, called the 'McCarren Committee Report'.

Historians may try to convince us that the communists came to power in China, when Chairman Moa's victorious revolutionary army overpowered the Nationalist forces of General Chiang Kai-Shek. But this could not be further from the truth. In fact, it was the pro-democratic Nationalist forces who were comfortably defeating the communists throughout the country.

Only when the US State Department stepped in and

embargoed Chiang Kai-Shek's weapon supply, did the communists turn the tide. Factions from within the US government placed an international embargo on the Nationalists, so they couldn't obtain weapons from anywhere in the world. Even blockading the weapons they had already purchased.

Taken from Dr Stanley Monteith's book, *Brotherhood of Darkness*, an extract from the 'McCarren Committee Report' reveals the reason why China fell to the communists.

> *"At the end of 1945 when [US Army] General Marshall left for China, the balance of power was with the [pro-democratic] Chinese Nationalists… and remained so until at least June 1946… Chiang's divisions were chasing the communists northward and the prospect of victory by Nationalist China was at its highest… However, when General Marshall arrived in China, he undertook to bring about the coalition government which his directive demanded… This plan failed when coalition failed… When the Chinese government did not effect coalition, by the summer of 1946 United States military assistance to China was brought to an end. Not only did the United States stop sending military supplies to the Chinese Government; the shipment of war materials actually purchased by the Chinese also was halted… The Chinese also had purchased surplus equipment that remained on Okinawa and other Pacific Islands. Even the shipment of this was banned… A complete*

embargo took effect in the summer of 1946. It was maintained at least until May 1947. General Chennault testified that the first shipment arrived in Shanghai in December 1948... Chennault further stated that the war material sent to China after the embargo did not arrive in time to aid the Chinese Nationalists in the field... Admiral Cooke... testified that the Chinese had a number of divisions equipped with American arms... When the flow of American ammunition was stopped, these divisions lost their fire power and were defeated. Even after the Eightieth Congress appropriated $125,000 000 for aid to the Chinese [Nationalists], shipments were delayed and when the guns finally reached the Chinese general in North China they were without bolts and therefore useless."

**–Institute of Pacific Relations Report
of the Committee on the Judiciary
Eighty-Second Congress
Second Session
S.(enate) Res.(olution) 366
A Resolution Relating to the Internal Security
of the United States**

According to the testimonies of General Chennault and Admiral Cooke, the Nationalist forces were starved of vital weapons and ammunition. So, it appears that without the help of factions from within the US government, the communists could never have taken China. And as we will see in the next section, those US government factions are controlled by the *financial elite*.

Finally, when it comes to wicked amoral indifference, almost nothing compares to the Chinese cultural revolution. With an estimated death toll of somewhere between fifty and one hundred million victims, it's testament to the all-consuming potential evil permeating within man's heart.

So, it seems the *financial elite*, rule over both Russia and China through their intelligence systems.

If that is the case, then surely those same people must also be in control of western intelligence. Interestingly, there does appear to be evidence connecting the *financial elite* to western intelligence, and strangely enough, part of it involves a little-known bank, as well as the importation of Nazis after WWII.

WESTERN INTELLIGENCE

It is common knowledge that the United States has the most powerful and well-funded intelligence agency in the world, but what is not widely known is how it was created, and who really controls it.

Ever since the American Revolution, the United States government has employed different types of intelligence gathering methods, but they were formed during times of war, and then disbanded. It wasn't until the outbreak of WWI that the government began building a world-class intelligence apparatus.

In July 1941, just before the US officially entered WWII,

the Office of the Co-ordination of Information was formed, which was then reformed into the Office of Strategic Services (OSS) in 1942. After the dissolution of the OSS in 1945, the SSU was created but only lasted for several months. Then in 1947, President Harry S. Truman created the first true peacetime intelligence apparatus when he signed the 'National Security Act' into being. Among other things, the Act created three brand new government entities.

- National Security Council.
- Department of Defence (Unifying Army, Navy & Airforce).
- Central Intelligence Agency (CIA).

The National Security Council is seen as the core foreign policymaker for the United States government, and that's not all, it also appears to have multiple tiers of power. While the number of core members changes from one administration to another, there are usually several 'statutory' and 'non-statutory' members. The top 'statutory' positions are currently the President, Vice President, Secretary of State, the Secretary of Energy, and the Secretary of Defence. Intelligence agencies, which include the CIA are represented by the Director of Intelligence (DNI), but it is also known that the CIA has very close links to the State Department.

It is this small group of people, plus their advisors from the intelligence community, who are at the heart of the United States government. But what is even more revealing is the fact that many of the 'statutory' members of the NSC since WWII, have also been members of the privately run Council on Foreign Relations (CFR), and later on the Trilateral Commission. These secretive organizations are part of a global network that is loosely known as the Round Table Groups, and their inner workings will be described in greater detail in *World Government, by Consent* (chapter 14).

Simply put, CFR members are placed inside the US government by the *financial elite*, so they can implement their hidden agendas. A quote from former CFR member, Rear Admiral Chester Ward, uncovers the close link the secretive organization has with Wall Street.

> *"The most powerful clique in the elitist groups have one objective in common—they want to bring about the surrender of the national independence of the United States.*
>
> *A second clique of international members in the C.F.R. comprises the Wall Street international bankers and their key agents. Primarily, they want the world banking monopoly from whatever power ends up in the control of global government."*

–Rear Admiral Chester Ward

Famous Wall Street banking names like Rockefeller, Morgan and Warburg are the clique Chester Ward was referring to. High ranking people from the banking and corporate worlds litter the inner-core of the CFR, and from there they are sent into the halls of political power. A quote from former CFR member, Richard Barnet, states how this same hidden hand also controls US intelligence, which is the heart of western intelligence.

"Failure to be asked to be a member of the 'Council' [on Foreign Relations] has been regarded for a generation as a presumption of unsuitability for high office in the national security bureaucracy."

–Richard Barnet, 1972

The CFR is the intermediary between the international bankers and the US government. Furthermore, the same people who control the US government are also behind the scenes controlling the Russian and Chinese governments. It must be understood that the vast majority of spies and officials from the East and West hold genuine animosity towards each other, but at the highest levels of world finance, both sides are controlled.

When the link between the East and West is established, it becomes easier to see who controls the highly classified Tesla technologies, and how they are being used against the nations of the world.

FASCIST INFILTRATION

When it comes to the ideology of American Intelligence, it is assumed by most westerners that its mandate is to protect democracy and free enterprise throughout the world. But if you look into the history of the Office of Strategic Services (1942-1945), Strategic Services Unit (lasted six months), and Central Intelligence Agency (1947 onward), then you will begin to see a set of secretive organizations full of treason and deceit, and that many nations who were once prosperous and peaceful have been left ruined in their wake.

Before the CIA was created, its predecessors, the OSS/SSU, are known to have had intelligence links to the Nazi regime. One of those connections was through a man called Thomas McKittrick. He was the head of a bank called the Bank for International Settlements (BIS) in Basel, Switzerland. This little-known bank was the institution responsible for providing foreign capital to the Nazi war machine while receiving vast hoards of war booty from conquered countries.

It would have been treasonous for any American to be seen dealing with the Nazis, but the relationship went much deeper than just spying or subversion. In an investigative article for the *Tablet*, author, Adam LeBor speaks of the close relationship the OSS had with the Nazis, and the roll McKittrick played in fostering it.

> *"As head of the BIS, headquartered in Basel, from 1940 to 1946, McKittrick played a crucial role in abetting Hitler's war—and, at the same time, in revealing details about his Nazi colleagues to his friends in Washington, DC.*
>
> *On McKittrick's watch, the BIS willingly accepted looted Nazi gold, carried out foreign exchange deals for the Reichsbank, and recognized the Nazi invasion and annexation of conquered countries. By doing so, it also legitimized the role of the national banks in the occupied countries in appropriating Jewish-owned assets. Indeed, the BIS was so*

indispensable to the overall Nazi project that the vice-president of the Reichsbank, Emil Puhl—who was later tried for war crimes—once referred to the BIS as the Reichsbank's only "foreign branch". In the closing months of the war, as American GIs fought their way across Europe, McKittrick was arranging deals with Nazi industrialists to guarantee their profits after Allied victory.

But McKittrick was also a key contact between the Allies and the Nazis, passing information back and forth from Washington to Berlin. His relationship with the Third Reich was encouraged both by factions within the State Department and by the leadership of the Office for Strategic Service, the predecessor of the Central Intelligence Agency."

–Adam LeBor
Tablet,
30th August, 2013

It appears that without the BIS-Reichsbank (German Central Bank) connection, it would have been impossible for the Nazis to finance military expansion the way they did. While American and Allied soldiers were dying on the fields of Europe, the OSS through McKittrick, were helping the Nazis pay their bills, and pillage the continent's wealth at the same time.

It has been suggested that treasonous people like McKittrick were nothing more than amoral opportunists, but when it is revealed how the OSS/SSU/CIA helped thousands of Nazi war criminals escape trial, then it becomes clear there was more than opportunism involved, but a close ideological connection appears to be present. That poorly understood episode involves the deliberate protection and importation of Nazi specialists and officials during Operation Overcast and later Operation Paperclip. Unlike other German leaders who were facing trial in Nuremberg, these Nazis were given special treatment by US intelligence. Some of them were even kept in high-class European hotels before being transferred to brand new state of the art towns in the United States. Three examples below, expose the moral bankruptcy exhibited by US intelligence officials.

- Arthur Rudolph–Chief Operations Director at Nordhausen, where 20,000 slave labourers died producing V-2 missiles. Described as a fervent Nazi, he led the team that built NASA's Saturn-V rocket.
- Wernher Von Braun–the SS Officer and rocket specialist who helped design missiles that bombed civilians in English cities. He was taken to the US, where he helped lead the space program, and was even made into an American hero.
- Hubertus Strughold–used by NASA for designing the onboard life-support systems. His subordinates conducted experiments at Dachau and Auschwitz, where some inmates were frozen to death, or placed in low-pressure chambers, where many died.

American intelligence officials knew full well that many of these men should have been sent on a one way trip to Nuremberg, but because of their expertise, they were given automatic exoneration.

But it wasn't only men like Rudolph, Von Braun, and Strughold who made their way into influential positions. When it came to running some of Americas most secret military programs, a veritable who's who of Nazi war criminals were involved. Below are four examples:

- Ernst Stulinger–climbed to the position of Director of Advanced Research Projects Division, in the Army Ballistic Missile Agency.
- Eberhard Rees–rose to the position of Director of the Marshall Space Flight Center.
- Kurt Heinrich Debus–went on to became Director of the Kennedy Space Center.
- Walter Dornberger–eventually became the vice president of Bell Aircraft Corporation.

The last example is extremely interesting, since Walter Dornberger oversaw electric propulsion research at Bell Aircraft. While some have suggested his research involved nothing more than electric fan engines, a 1955 report from *Aviation Studies* states that Bell Aircraft, along with many other aerospace companies were working on unconventional aircraft. Even Dornberger himself had stated that passenger aircraft would eventually fly at 10,000 miles an hour (Mach 13), making the trip from New York to Sydney in a leisurely one hour. If this is correct, then it gives credibility to the argument that Paperclip-Nazis were at the very heart of Americas 'deep black' military programs, and the rumours surrounding super advanced jet engines or even Anti-Gravity research may, in fact, be correct.

It appears the relationship between the Nazis goes much deeper than just the transfer of advanced technology, there is also evidence suggesting US intelligence was also after their ideology. The proof for this came in 1998, when the US government passed the 'Nazi War Crimes Disclosure Act', in an attempt to uncover the whereabouts of the Paperclip-Nazis, and how far their ideology had infiltrated intelligence circles.

Because of incessant stonewalling by the CIA, the Act was only partially successful in uncovering the truth. One interesting case that did come to light was of a General by the name of Reinhard Gehlan. He was Hitler's Chief of Intelligence in the German-occupied eastern front. It was Gehlan's intelligence that was critical in implementing the Nazi death squads, which saw millions of Russians, Gypsies and Jews sent to their deaths. The excuse for recruiting Gehlan, and many of his henchmen into the Strategic Services Unit (SSU), was because it was seen as a vital tool in fighting the spread of communism. That's because the Nazis knew how to gather intelligence on the ground, and there was no moral compunction as to how that information was gathered.

It is believed the deal struck between the Nazis and the US government wasn't one of master and servant, or even the joining of equals, but was, in fact, an ideological takeover. This can be seen when Gehlan was able to recruit thousands of Gestapo, Wehrmacht and SS veterans into his organization, thus allowing them to create a large yet secretive clique within the agency. Based on information from the 'Nazi War Crimes Disclosure Act', the points below reveal the evil intent of Gehlan's group.

- Gehlan operatives took leading roles in the creation of new European fascist organizations.
- Gehlan's neo-fascist organisation proved to be an effective backdoor for Soviet infiltration of the western alliance.
- Members of the Gehlan organization helped thousands of Nazis flee Europe through the 'ratlines'. Some of those fugitives wound up security advisors in several Middle Eastern and Latin American countries. They helped with torture, as well as run the drug trade and death squads.

It has been suggested by some researchers that the CIA just used a 'nod and a wink' attitude concerning Gehlan's atrocious conduct. But the facts tell a different story. For a start, Gehlan was fully financed by the CIA, and in certain quarters of the agency he was highly regarded for his expertise. If that is the case, then surely they knew what he was doing. In *World Government, by Conquest* (chapter 15), it describes some of the nefarious activities the CIA is conducting today, and how that same ideology has continued into the twenty-first century.

If this insidious intelligence program was to work, then secrecy was the key to keeping it operational, and because much of what went on was totally uncivilised, it may have been necessary to even keep Presidents out of the loop. Annie Jacobson, the author of *Operation Paperclip—The Secret Intelligence Program, that Brought Nazi Scientists to America*, revealed the level of secrecy imposed by those who control these programs. In an interview, she states how President Ronald Reagan had tried to find information on a notorious Paperclip-Nazi called Otto Ambros, but even he, as a sitting US President, was denied access. Reagan was a fervent anti-communist, so for the *financial elite*, it would have been important to keep elected officials like him in the dark, which just goes to show—the President is not privy to everything. Conversely, others in Reagan's Administration went out of their way to stop the Paperclip-Nazis from being investigated by the Justice Department.

While fascist infiltration of US intelligence was well underway in America, the Soviet Union was already under socialist rule, so it wasn't hard to incorporate Nazi ideology into their intelligence apparatus. It is now known that they too swooped on the Nazis, in what became known as Operation Osoaviakhim, a program which saw thousands of officials and scientists, as well as their families, transferred to the communist country. While many believe the Nazis were taken to the Soviet Union by force, this may not, in fact, be the case, according to authors P. Jackson and J. Siegel, in their book *Intelligence and Statecraft*, many of the Germans were lured with large salaries and plenty of food and fuel. And for both Soviet and American Nazis alike, there appeared to be a degree of freedom (including travel) that you would not expect for captured war criminals.

It has been suggested by some researchers that the United States was the victim of a soft coup when the 'National Security Act' was introduced in 1947. From the information gathered here, I would be hard-pressed to disagree with that statement. Also, we must assume from the information in this segment, that the Nazis, the Communists and the *financial elite* share the same ideology, the only difference is, the latter has gone to great lengths to keep their fascist beliefs secret. Even after seventy years, they still successfully maintain control of both the eastern and western intelligence systems, and with that the highly classified technologies.

CONTROL OF TESLA'S SECRETS

In the early nineteen-eighties, some of Tesla's FBI files were declassified. They reveal how some of the inventor's papers weren't released to his native Yugoslavia, specifically his work into ball lightning, and its use for aircraft propulsion.

Also, as it was previously described in *Wardenclyffe* (chapter 6), when US intelligence released his remaining papers, it was assumed by the FBI that there were no breakthrough discoveries left to extract. But according to author and researcher Tom Bearden, the Soviets were fully aware of Tesla's Aetheric (scalar wave) experiments, and were ready to glean any information they might contain. He believes the Soviets were quick to develop Tesla's wireless power into a weapon, and that they are not used by the regular Soviet (Russian) armed forces, but are tightly controlled by the KGB/FSB.

No other weapon in existence can possibly win against the Aether superweapons, so again it raises the question from earlier in the chapter: if the Soviets were in possession of them, then why didn't they use it to destroy the West? The answer is simple, the super advanced technology programs in both the East and West are controlled by the same people, and regardless of which country they are situated in, they are being used for a common purpose.

This explains why both the Bell Island event, which

involved Americans, and the Soviet-based Hindu Kush incident are covered up by both sides. Why else would either superpower restrain their use in an all-out war, when it is clear a surprise attack would cripple their opponent.

The super advanced technologies are ultimately controlled by the *financial elite*, the same people who built and nurtured the socialist movements. They have spent vast quantities of time and money developing these breakthroughs, some of which have the potential to free mankind from many of the ills that plague it, but instead of freedom, all they wish to do is use it to bring about global slavery.

While advanced technologies are extremely important to them, they are only one part of a multi-faceted plan that ties together people from the military, politics, religion, intelligence, and most importantly finance. So who really are the *financial elite*? They are an important part of a hidden cabal, who have spent centuries trying to bring the wealth of the world under their control, and the financial system they have created for directing their affairs is nothing short of startling.

The following chapters will uncover the inner workings of that monetary system, and how they have used it to financially colonize the world. Then it will attempt to uncover the blueprint of how they plan to introduce their ultimate goal, a 'One World Socialist Government'.

CHAPTER TWELVE

THE CITY

"We should be clear about this: the City of London is now being viewed by many as a tax haven in the middle of a dense network of havens created for the super-rich to avoid the taxes the rest of us must pay."

–Belfast Telegraph Digital, 13th April, 2016

FOR ANY WORLDWIDE conspiracy to be successful, it must establish itself through three main avenues. Firstly, the conspirators must gain control of the nation-states through subversive or military means. Secondly, they must establish a global culture through a new type of ideology or religion. But most importantly, they must take full control of the world's financial system because, without it, they won't be able to implement any of the other globalist plans.

So, for those who desired to control the world, the realization would have emerged quite early that they needed access to vast reserves of the world's wealth.

Throughout the centuries certain commercial centres

of power were just as influential if not more so, than the political capital they came under. Looking back throughout history from the oldest to the newest, some of those international cities are mentioned below:

- Ur—Ancient Babylon (Sumer)
- Nippur—Ancient Media-Persia
- Piraeus—Ancient Greece
- Ostia —Ancient Rome
- Venice—City-State (Maritime republic)
- Genoa—City-State (Maritime republic)
- Antwerp—Duchy of Brabant
- Amsterdam—Home of the Dutch East India Company
- City of London—Home of the British East India Company

It is believed by most historians, that the foundations of the financial system in operation today had their origins in the Babylonian Empire. For example, loans were known to have existed between agricultural merchants and traders as far back as 2000 BC, and a form of banking goes back to the Code of Hammurabi.

Over the millennia, various aspects of lending have been refined, but the religious temples remained the central location for the practice. Then in around the 12th century, powerful City-States like Venice and Florence took banking from its Temple based system to the more modern set-up we see today. And from that point on, powerful families such as Bardi, Peruzzi and Medici, took the new banking system out across Europe, eventually giving rise to some of the richest city-states in history.

CITY OF LONDON

Some of the information contained in the next two sections are collated from investigative articles in The Guardian, *as well as from Nicholas Shaxson, the author of* Treasure Islands.

Generally accepted as being established by the Romans, Londinium quickly became an important trading outpost for the Empire. By the 4th century, stone walls had been built to protect it from possible uprisings, helping to establish it as the prominent seat of power in Roman-controlled Britain.

By around the end of the 5th century, Rome had left Britain, and for the next three hundred years, the walled town became mostly uninhabited. Then by the 9th century, Alfred the Great (871–899) began to resettle the fortified ruins. It is believed that at the time of Edward the Confessor (1042–1066) and William the Conqueror (1066–1087), special rights and privileges began to be bestowed upon the citizens of London, that were not imparted to the rest of the population. Then gradually over the centuries, further concessions had been obtained from the Crown.

As the financial might of the City began to grow, it became a beneficial relationship for both sides. The Crown could borrow money to do things, like extend land ownership, wage wars, or implement social projects. Likewise, the City could rely on the Crown to honour its special charter. But it hasn't always gone the way of the City. Throughout its history, monarchs and members of parliament alike have tried to abolish its special privileges, but to this day they have been unable to do so.

To most people around the world, the modern metropolitan area known as Greater London (8,500,000 people) is nothing more than the political, cultural and financial hub of the United Kingdom. But the reality is quite different. Inside Greater London on the north bank of the Thames sits the old City of London (12,000 people). Because of its size, some have termed it the Square Mile, but for the sake of simplicity, the term Square Mile will be used to describe the physical land, while the City will pertain to the body that runs it.

So, what makes this city so different? Even though the City has a Lord Mayor, it is not run like any other city or town in the United Kingdom. To start with, there are twenty-five electoral wards, each having an Alderman. Those twenty-five wards also have Councilmen who make policy for the City of London. Where the City differentiates itself from other municipalities is when it comes time to vote. Of the twenty-five wards, only four have residential voting, the other twenty-one, allow voting by companies that reside within the enclave. Also, if the Aldermen don't like the person elected, they can use a law from 1252 AD, to rescind the elected official's position. So, unlike all other councils in the UK, companies in the City can vote in the electoral wards, and not just that, their votes are based on the size of their workforce. For example, ten employees get one vote, while fifty employees get two votes, and so on, and all this without the knowledge of their employees. It is a system much older than Westminster itself and is believed to be the oldest municipal democracy in the world still operating.

Another important privilege the City enjoys is that

it has its own special representative in parliament called The Remembrancer. That official sits in the House of Commons and can veto or get special dispensation from any legislation that may affect its financial interests. It may be a part of the UK and enjoy all the privileges that brings, but it operates like a tax haven with very little governmental oversight.

This little enclave operates under archaic feudal privileges and that's why it's called the City of London Corporation. Yet its influence doesn't stop at the borders of the UK but extends out like a spider's web to all corners of the globe.

OTHER FACTS ABOUT THE CITY OF LONDON

- Boasts 41% of global foreign exchange trading.
- 85% of European based hedge fund assets.
- $330 billion of worldwide premium insurance income.
- There are around 500 banks operating in the City of London, many of them foreign.
- Location of Lloydes insurance/reinsurance Marketplace.
- Location of the London Bullion Market Association (LBMA). Sets the daily spot price for gold, silver, platinum and palladium around the world.
- Location of the London Metals Exchange (LME). The world centre for industrial metals trading.
- Home of the Bank of England.
- Home of the London Stock Exchange.

TAX HAVENS

When talking about tax havens, it's easy to think of places like Switzerland or Monaco. But deeper research into this clandestine area of global finance reveals a startling truth, and that is the City of London, along with many of the former British colonies, have set up a web of finance that is unequalled anywhere in the world. The British Empire officially ended around the time of the Second World War, but because many of its small islands lacked size and population, it was impossible for them to attain financial independence. That forced many of them to either stay under full British control (Crown Dependencies) or become partially self-governed as British Overseas Territories (BOTs). It is from this background that powerful financial interests within the City began to set up tax havens in various locations.

According to Nicholas Shaxson the author of *Treasure Islands,* if the City of London is the core, then the first layer of tax havens are the British Islands of Jersey, Guernsey and the Isle of Man. The second layer is the British Overseas Territories (BOTs) like Bermuda, British Virgin Islands, Cayman Islands, Turks and Caicos Islands, and Gibraltar. The third layer includes cities like Hong Kong, Dubai, Nassau, Singapore and others. It's these three tiers that provide the bulk of the wealth that gets funnelled back to institutions in the City of London.

An example of just how much wealth flows into the City from outlying tax havens, can be seen in the 332.5 billion (US) dollars that moved from Jersey, Guernsey and the Isle of Man, in just three months of 2009. Some estimates suggest that this global web of untaxed and unregulated wealth gives those who control it just over half of all the world's offshore deposits, which equates to tens of trillions of dollars.

Another interesting aspect about the City and her web of tax havens is its ability to launder money. Many who have investigated this area believe it is the heart of the global money laundering business, with terrorists, despots, drug kingpins, and warlords the world over washing their ill-gotten gains inside tax havens and trans-national banks. One interesting case, among many, was when British banking giant HSBC was caught laundering money for the Mexican drug cartels, as well as various terrorist organizations like al-Qaida, in 2012. The company was fined 1.9 billion (US) dollars, which may seem adequate for most of us, but was just five weeks income for its American division alone. It doesn't take long to realize that it was nothing more than a slap on the wrist for a company that has a higher income than most countries.

Banks aren't alone in this; the famous Trans-National Corporations (TNCs) that operate around the globe, also hide their vast profits. While estimates vary, one report released by *Citizens for Tax Justice* (*CTJ*) states that 358 of the Fortune 500 companies held more than 2.1 trillion (US) dollars offshore for tax purposes. If that money was sent back to the United States, the tax bill could be as high as 600 billion (US) dollars. Furthermore, according to author Nicholas Shaxson, the US government misses out on 100 billion dollars in annual taxes because of this system. By simply setting up a subsidiary in somewhere like the Cayman Islands these TNCs can route large portions of their income into what is essentially a Post Office Box, with a corporate deed sitting inside. It may be a practice that is legal, but it is immoral and devoid of social empathy. While the lower and middle classes are being burdened with higher taxes, the elite are being taxed next to nothing.

The tax haven might be one of the greatest threats to middle-class free enterprise and true democracy, but that is only one layer of the deceit; another one involves the way big companies operate. When it comes to the largest TNCs in the world, for them 'competition' is a dirty word, and the plan they came up with to counter it, is nothing short of staggering.

TRANS-NATIONAL CORPORATIONS

As large companies grew throughout the twentieth century, especially with the help of the ruthless robber barons, it also appears there has been a plan to unify these already bloated companies into a single web of corporate power. Many accusations in the past have implied that a powerful cabal influences the corporate world from behind the scenes, but it has been a hard accusation to prove. Then in 2011, an article in *New Scientist* describes what theorists at the Swiss Federal Institute of Technology in Zurich found while studying mathematical models of natural systems. By combining their mathematical models with comprehensive corporate data, like shareholdings, shared directorships, and indirect ownership, they discovered that out of the 43,060 Trans-National Corporations (TNCs) that operate globally, there appeared to be a hidden web of closely linked companies. From the article in *New Scientist*, a picture begins to emerge, describing how this web operates.

"Each of the 1,318 (companies) had ties to two or more other companies, and on average they were

connected to twenty. What's more, although they represented 20 percent of global operating revenue, the 1,318 appeared to collectively own through their shares the majority of the world's large blue-chip and manufacturing firms—the 'real' economy—representing a further 60 percent of global revenue."

– *New Scientist*

When the team further untangled the web of ownership, it found much of it tracked back to a 'super-entity' of 147 even more tightly knit companies.

Top twenty of the 147 super connected companies.

- Barclays Bank PLC
- Capital Group Companies, Inc
- FMR Corporation
- AXA SA
- State Street Corporation
- J.P. Morgan Chase & Co
- Legal & General Group PLC
- Vanguard Group, Inc
- UBS AG
- Merrill Lynch & Co, Inc
- Wellington Management C.O., LLP
- Deutsche Bank AG
- Franklin Resources, Inc
- Credit Suisse Group
- Walton Enterprises, LLC

- Bank of New York Mellon
- Natixis SA
- Goldman Sachs Group, Inc
- T. Rowe Prize Group, Inc
- Legg Mason, Inc

When the Swiss team assessed the information, they discovered a web of 1,318 TNC's that they considered 'very-connected', but that wasn't all, they also found a core of 147 'super-connected' corporations at its heart; this they dubbed the 'super-entity'. Overall, about 40% of all the worlds TNCs. are connected to the 'super-entity' in some way. Also, it is believed that the core of companies (banks, investment houses, insurers, etc) at the heart of this web control most of the famous blue-chip and manufacturing companies, equating to tens of trillions of dollars in value. While many people still consider New York as the world's leading financial centre, with many blue-chip companies headquartered there, it appears they are just part of the 'super-entity' that operates globally through the tax haven system.

One thing that is very interesting is how those who conducted the experiment did not come at the study from a conspiratorial point of view, they were just trying to establish how big companies were structured. To best understand how the secretive corporate 'super-entity' works, one needs to look at the non-profit organizations that hold a lot of the stock in these corporate giants. A quote from Gary Allen's book *The Rockefeller File*, best describes how this magical 'sleight of hand' takes place.

"The Rockefellers invented a scheme, used by
the super-rich today, whereby the more money

you appear to give away, the richer and more powerful you become. Through the help of captive politicians, guided by some bright boys in the family law offices, legislation was written and passed which would protect the Rockefellers and other elite super-rich from the repressive taxation they have foisted on every-one else.

The key to this system is giving up ownership but retaining control. For example, most people don't believe they really own something unless they retain title to it in their own name. The Rockefellers know this is a big mistake. Often it is better to have your assets owned by a trust or a foundation—which you control—than have them in your own name."

–Gary Allen
The Rockefeller File

Own nothing and control everything, that is the motto for the secretive elite who want to control world finance. By placing the shareholdings from all their corporate interests, behind a wall of hundreds, if not thousands of shell companies, trusts and foundations, the elite (not just the Rockefellers) can hide a company's true ownership, and all through the loophole of philanthropy. Giving away some of their wealth through 'charitable' donations from their trusts paints the globalists with a veneer of altruism, sold to the public through fawning press coverage in the tame mainstream media, who in turn are subsidiaries of the elite.

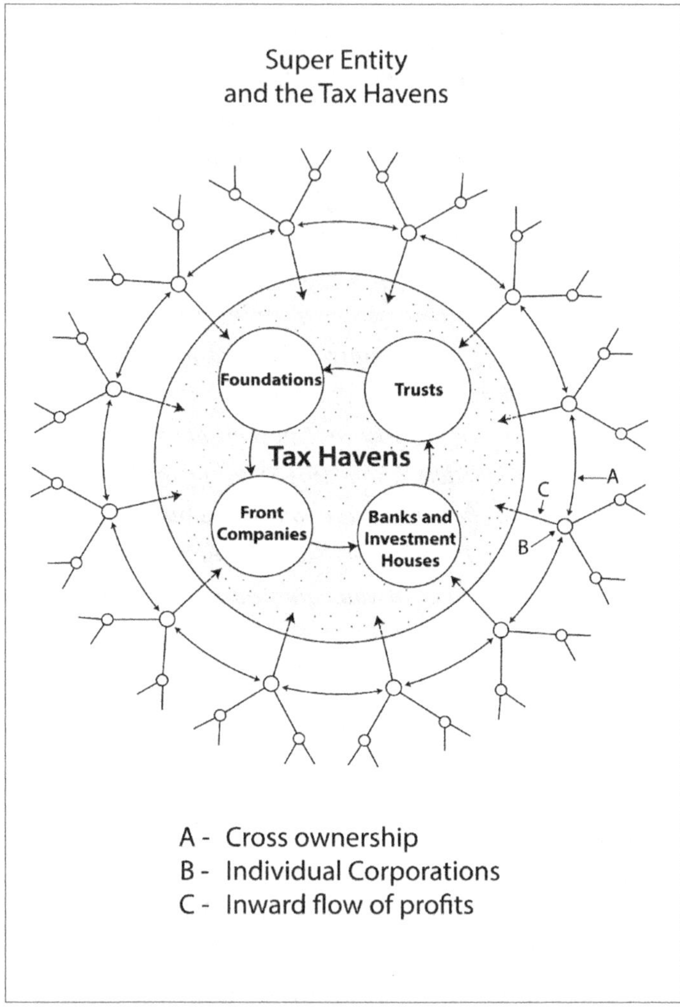

Super Entity and the Tax Havens

The tax havens are a powerful tool in the hands of the *financial elite*. Money flowing in from the Trans-National Corporations and banks, can be moved around the world with very little government oversight. While many well-to-do people use tax havens, they are primarily used by the elite for the purpose of pooling much of the worlds wealth, and then using it for various purposes, including the cultivation of political influence around the world

Another benefit that type of arrangement allows, is in the case of litigation. If one of their companies commits a serious offence, the owners can deny any culpability, while the board of directors take all the flack. This is exactly what happened when the giant agrichemical company Monsanto was hauled before the courts for illegally influencing governments over the introduction of Genetically Modified Organisms (GMO). In a rare lapse of media censorship, the negative press got so bad for Monsanto that one article in the *New York Times* stated how the board of directors were being hauled before the Rockefeller Foundation (their true owners), and handed a "Please Explain".

When it comes to the banks and investment houses that control the world's large TNC's, it appears all roads lead back to the City of London. These banks may have headquarters and foreign offices around the world, but many hundreds have offices situated inside the City. It could be suggested that these branches are nothing more than backwater operations, but that doesn't appear to be the case, as stated here by one of the big Japanese banks:

> *"All large Japanese city banks have London branches, belonging to Japanese banks in Japan. Therefore, they do not need to produce annual reports of their activities in London. Usually the managing directors are global members of the Board of Directors in Japan, though they appear to hold much higher positions than they had in Japan."*

> **–Junko Sakai**
> ***Extract from Japanese Bankers in the City of London***

Large banking houses from around the world operate branch offices in the City of London, and they are run by high-level board members. And surprisingly enough, it isn't only Western banks with representation in the City, but communist lending institutions are there as well. One example is Moscow Narodny Bank (MNB). It was set up just prior to the 1917 Revolution, and acted as the financial foreign office, so the Soviets could interact with powerful western financiers. This east/west relationship also extended to China, with the Bank of China (BOC) setting up a City branch in 1929. Then after the communists took power in 1949, the London branch became the centre for debt management, foreign exchange and international trade. Even today, with China still officially communist, her five 'big' banks now house foreign offices in the City.

Why would this so-called bastion of capitalism, want to interact and bolster communist regimes? The answer is simple, the City of London is the centre of world finance, for both the capitalist, as well as the communist worlds. It is also the enclave where the visible and shadow financial worlds interlock, giving those who control it the power to influence not only the financial realm but the political domain as well.

CHAPTER THIRTEEN

THE CENTRAL BANKERS

After the first central bank failed, another attempt was made in 1816, the second bank also failed. This time it was President Andrew Jackson who refused to sign its second twenty year charter. Such was the President's enmity towards the bank that his electoral slogan was 'Jackson and no bank'.

–A short summary describing the battle between President Andrew Jackson and the second central bank of the United States.

THE FOLLOWING SECTION describes the relationship between a central bank, its government, and the private banking sector.

Most people have heard the term "central bank", but they wouldn't be able to adequately explain its purpose, or how much control it wields over the economy of its respective nation. Officially, the job of the central bank is to help stabilize the economy, by setting interest rates on loans and mortgages, as well as regulate the private banking and insurance industry. Another important job it does is manage the nations debt (through Treasury Bonds, Bills and Notes), so the government can operate a full budget while lacking adequate taxes.

The story of who controls central banking is still to some extent shrouded in mystery. Most say it is their respective governments, but when one studies their history a different picture begins to emerge. The United Kingdom may not have been the first country to introduce a central bank, but it is believed to be where the central banking system as we know it today, had its origins.

Established in the City of London in 1694, and otherwise known as 'The Old Lady Of Threadneedle Street', the Bank of England was charged with one important task no other banking institution had. That is, it received a Royal Charter from the government to supply all the credit the Treasury would ever need to run their programs. More importantly, it was a group of private investors who received the Royal Charter, so in reality, the Bank of England is a Public-Private-Partnership (P.P.P.), but as we will see later in the chapter, the 'private' element appears to be the dominant partner.

What about the credit creating powers of the private and central banks? According to William Paterson, the Scotsman who helped create the Bank of England, it is their ability to be able to bring money into being with nothing more than the stroke of a pen, that gives them such power.

"The bank hath benefit of interest on all moneys which it creates out of nothing."

*–William Paterson, 1694., requoted in 1907
by the Chairman of the Midland Bank.
(Accepted by most historians)*

Called "Bank Credit Creation", its very existence was admitted by the Bank of England in 2014.

Essentially, what it means is, a private bank can create new money out of nowhere when it provides a new loan for a borrower. In other words, the bank does not have to supply the loan from deposits it receives from its customers. When it's understood that 97% of all money in circulation is credit (3% is physical cash), then it becomes clear that private banks hold immense power over the financial system. So what does this have to do with the central bank? Because the central bank oversees the whole banking system, it has the ability to use debt-creation for controlling the entire economy.

According to economist and author Richard Werner, another important function central banks perform is colloquially known as 'window guidance'. In short, 'window guidance' is where a central bank allocates credit through the private banks for a specific area of the economy, like real estate, heavy industry or tech start-ups. In the nineteen-eighties, economists discovered that the Bank of Japan was using 'window guidance' to over-inflate the real estate market, thus causing one of the worst economic recessions in recent Japanese history. What this suggests is, that central banks have the power to be able to inject credit into or withdraw it out of, any part of the economy, and in doing so they can create bubbles, recessions and depressions. According to some economists, evidence suggests that

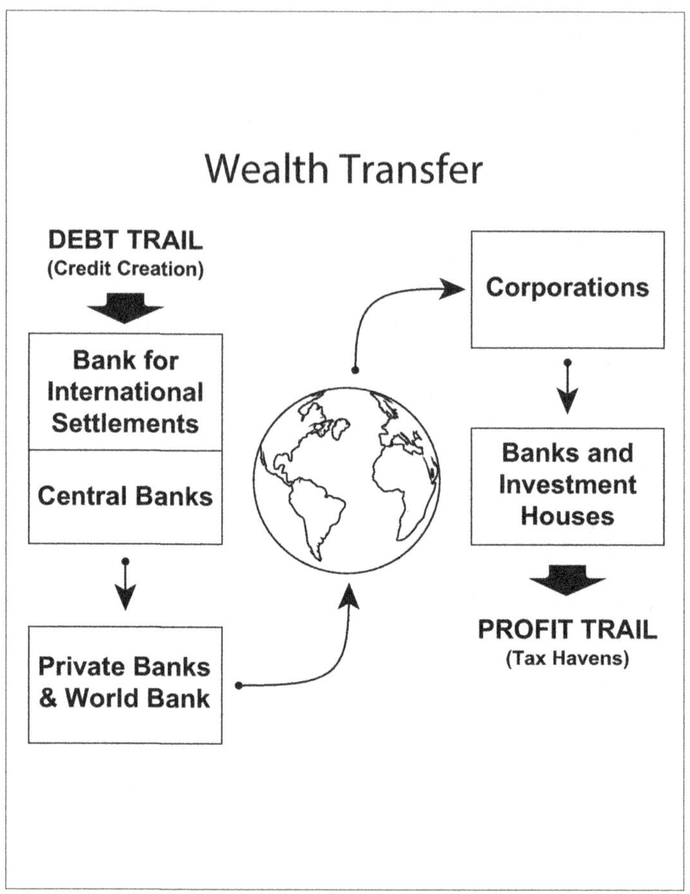

Wealth Transfer

For years the *financial elite* have been systematically transferring the
worlds wealth into their own hands, and the mechanism they use
to achieve this, is debt. By creating credit out of nowhere, private
and central banks are able to "in debt" everyone from individuals
to countries, while they use the same credit making power to buy
up all the profit-making assets.

boom/bust cycles are not the result of lifting or lowering interest rates, but are caused by the injection or retraction of credit by the central bank. If that is the case, then central banks have the power to impose poverty or prosperity upon society at the press of a button.

What about the loans provided to governments, how are they serviced? After the money is loaned out to the government at interest, the taxpayer, through their future taxes, service the loans, so in the end, it is a win-win for those who control the central bank. It's interesting to note that since the inception of the Bank of England, the United Kingdom has never been free of debt, and the same could be said for most other nations as well. Another very telling admission concerning the power private and central banks have, came from Reginald McKenna, the British Chancellor of the Exchequer and Chairman of Midland Bank.

"I am afraid the ordinary citizen will not like to be told that the banks can and do create money and they who control the credit of the nation direct the policy of governments and hold in the hollow of their hands the destiny of the people."

–Reginald McKenna,
1924

McKenna's statement reiterates Paterson's own quote, about the power the banking system has when it comes to creating money out of nowhere. By loaning large volumes of money to the nation, they are able to influence the direction of the government and private sector for whatever reason they choose.

When it comes to the Bank of England, she was officially nationalized in 1946, but this farcical act meant nothing; years later it was revealed that it remained independent from political interference and if anything, was given more power than ever before. Since W.W.II., the Bank of England has lost some of its international pre-eminence, but there is another central bank with size and scope that boggles the imagination. That bank is the Federal Reserve of the United States, and just as private businessmen helped create the Bank of England, the same kind of people were at the founding of the Federal Reserve.

U.S. FEDERAL RESERVE

From the time of the American Revolution, private bankers have strived to introduce a central bank, but barring two short periods, they were largely unsuccessful.

Fast forward several decades and the American people were growing weary of the economic instability throughout their country, mostly because it was leaving many small banks vulnerable to closure. What the people didn't realize was, the owners of the large private banks had created the instability in the first place. Their plan was simple: create the problem, then fix it by creating an agency to oversee the whole banking industry. It wasn't going to be easy though, the people were wary of central banks, and knew they wielded a lot of unwarranted power. But that didn't stop the bankers, they just kept telling the people they could solve the economic problems the nation was facing.

A new attempt was needed, and according to G. Edward Griffin, in his book *The Creature from Jekyll Island*, the Federal Reserve was conceived, in a secretive meeting on Jekyll Island, Georgia in 1910. Included in that meeting were:

- Abraham Andrew, Congressman, prominent in banking circles
- Frank Vanderlipp, President of the National City Bank of New York, representing the Rockefellers
- Henry Davidson, Senior Partner at J.P. Morgan
- Charles Norton, President of First National Bank of New York
- Paul Warburg, Partner in Kuhn Loeb & Co, representative of the Rothschilds of Europe
- Benjamin Strong Jr., J.P. Morgan's Bankers Trust Company

These men drew up the plan for the Federal Reserve, but it would be another three years before it could be implemented.

Then on the 22nd of December 1913, while much of Congress was away on an extended holiday, President Woodrow Wilson passed the 'Reserve Bank Act', paving the way for the creation of the biggest money trust on earth. To deceive the public, the bankers set up twelve regional reserve banks around the country, so people wouldn't think one organization was taking over the whole financial system. Little did they realize that this was the plan all along. While the government appointed the governors to the Federal Reserve Board, the private bankers took ownership of stock in the twelve Regional Reserve Banks.

It's important to mention again that, central banks are not private companies in the truest sense of the word, but are more like a partnership between a cartel of banks and the government. This is seen in the ownership model of the Federal Reserve; for example, the shares the private banks own cannot be on-sold like normal stocks, and they don't give the bigger banks special voting rights over the smaller ones. In simple terms, the Federal Reserve is a unique type of hybrid agency, and because of its Special Charter, the private bankers are able to use Bank Credit Creation to build a giant slush-fund.

So, while the government appointed Board of Governors exercises a certain amount of influence over the Fed, as we will see in the following quotes, the private bankers are able to override any decision making, particularly through the Regional Reserve Banks, like New York.

It's interesting to note that within two decades of its creation, the US was thrust into the worst economic depression it had ever known. The promise politicians made about the Fed's ability to save the small banks was nothing more than a lie. While the Federal Reserve looked on, many of the small private banks went under, paving the way for the wholesale takeover of the US economy by an elite group of banking and investment houses.

The following quotes are presented in a way that shows the effort the *financial elite* have placed in getting economic control over the United States. The first quote is from one of the conspirators who helped create the Federal Reserve.

Insider Frank Vanderlipp stated to a magazine that it

was of the utmost importance that the American people be kept in the dark concerning the meeting at Jekyll Island.

> *"I was as secretive—indeed, as furtive—as any conspirator... Discovery, we knew, simply must not happen, or else all our time and effort would be wasted. If it were to be exposed publicly that our particular group had got together and written a banking bill, that bill would have no chance whatsoever of passage by congress."*
>
> **–Frank Vanderlipp**
> **Saturday Evening Post,**
> **9th Feb, 1935**

The very essence of what a central bank does and how powerful it is must be kept from the general populace, which is why it operates under such secrecy. Even when central banks are forced to speak publicly, it is usually in economic gibberish, so no one, including most economists, can truly understand it.

One government official who gave intimate details about the Fed was congressman Wright Patman (D-TX) of Texas. The congressman chaired the United States House Committee on Financial Services which oversees the entire financial system (including the Fed) on behalf of the government. While on the Financial Services Committee he saw things most Americans don't get to see. This is one statement he made concerning the Federal Reserve.

"We have what is known as the Federal Reserve System. That system is not owned by the Government. Many people think that it is, because it says 'Federal Reserve'. It belongs to the private banks, private corporations. So we have farmed out to the Federal Reserve Banking System that is owned exclusively, wholly, 100 percent by the private banks—we have farmed out to them the privilege of issuing the government's money. If we were to take this privilege back from them, we could save the amount of money... in enormous interest charges."

**–United States Congressman, Wright Patman,
Speech to House of Representatives,
1941**

Congressman Wright Patman understood the workings of the Fed, he knew that if it could be abolished the US economy would be freed from its interest rates and draconian control. That is why he pushed for its abolition for more than twenty years.

Another government official who was a controversial figure amongst his peers was Senator Barry M. Goldwater. In his book *With No Apologies* he describes how the Federal Reserve is a law unto itself. Here is an extract from the former Senator's book.

"Most Americans have no real understanding of the operation of the international money lenders... The accounts of the Federal Reserve

System have never been audited. It operates outside the control of congress and... manipulates the credit of the United States."

–United States Senator, Barry M. Goldwater, 1979

When Senator Goldwater states "international money lenders", he is actually saying *financial elite*, the true controllers of the Fed. He also states that no one has ever been able to audit its books to see who the Federal Reserve has been lending too. The truth is, if the books were audited, the people involved would be tried for treason.

Yet, many people still believe the Federal Reserve is an agency under the full control of the US government, even stating how the United States Treasury Department directs policy over the powerful bank. But again, that is not the case, because events written about during the early part of the Reagan Presidency reveal the troubles the new Administration was having with the Fed. One article in the *New York Times* outlines the problem.

"President Reagan today raised the question of whether the Federal Reserve Board should be placed under the authority of the Treasury Secretary... He [Ronald Reagan] *did not elaborate, but the remark revived the sensitive issue of crimping the Federal Reserve's autonomous status in setting monetary policy and influencing interest rates, and raised the question whether the Administration might still be*

*considering exercising some sort of oversight
over the Federal Reserve."*

**—The New York Times,
18 September, 1982**

It is believed that, at the time, President Reagan had tried
to bring oversight to the Federal Reserve, but couldn't. It
appears to be an entity which has full autonomy in directing economic policy in the United States, and to this day
it has never allowed itself to be independently audited.

The next quote is important because it establishes how
the Fed interacts with its twelve regional reserve banks.
In 2009, the *Bloomberg* news agency entered a lawsuit
against the Federal Reserve, after the government bailed
out Wall Street corporations, following the 2008 financial crash. The lawsuit was not about money but instead
wanted the Fed to release information concerning how
the bailouts were structured. Here is a part of the press
release concerning the relationship between the Federal
Reserve, and the Regional Reserve Bank of New York.

*"The Fed Board of Governors contends that it
is separate from member institutions, including the Federal Reserve Bank of New York,
which runs most of the lending programs. Most
documents relevant to the Bloomberg suit are
at the New York Fed, which isn't subject to
F.O.I.A. (Freedom of Information Act) law,
according to the Central Bank. The board
of Governors has 231 pages of documents, to*

which it is denying access under an exemption for trade secrets. The board cannot seriously maintain that the New York Fed does not perform governmental functions and control information of interest to the public.

The Freedom of Information Act obliges federal agencies to make government documents available to the press and public. The Bloomberg lawsuits filed in New York doesn't seek money damages."

**–Bloomberg, Press Release,
April 16, 2009**

This extract from the *Bloomberg* press release, makes it clear that the Federal Reserve Board of Governors is nothing more than an oversight committee representing the US government. The real important decisions are made by the regional reserve banks, particularly New York.

The press release goes on to say that the Regional Reserve Bank of New York is not subject to the FOIA, and thus does not have to reveal information concerning any loans. What this is really saying is that the New York Fed is not under the control of the Federal Reserve Board of Governors, and like all regional reserve banks are separate entities controlled by the private banking houses.

The final piece of evidence that must be mentioned points to how the US government services its debt, and why officials don't like to talk about the burden it places on the economy.

Officially known as 'The Presidents Private Sector Survey

on Cost Control' (P.P.S.S.C.C.), but otherwise known as the 'Grace Committee Report'. It was written for President Ronald Reagan, and designed to uncover the vast waste and overspending throughout the United States government. One interesting part of the report reveals the level of importance placed upon the servicing of Federal debt.

> *"... 100 percent of* [Federal Income Tax] *collected is absorbed solely by interest on the Federal debt and by Federal Government contributions to transfer payments* [Social Security Administration]. *In other words, all individual income tax revenues are gone before one nickel is spent on the services which taxpayers expect from their Government... "*

—The Presidents Private Sector Survey on Cost Control,
12th January, 1984

As stated in the final report of the P.P.S.S.C.C., servicing the interest on the national debt is an important part of the overall Federal Budget. Essentially, what that means is, all debt must be serviced before all the various agencies receive Federal funding. If future funding for those agencies cannot be fully met, then more loans must be acquired, and thus the debt just gets deeper and deeper by the day.

The mysterious world of central banking leaves those who study it without any doubt that they are controlled by elite private bankers, and their influence flows down into the political arena through the use of debt creation. But the story of central banking doesn't quite end there, because an even more mysterious organization lies at the heart of the world central banking system, that organization is called the Bank for International Settlements, or BIS.

BANK FOR INTERNATIONAL SETTLEMENTS

This section is largely based around the research of Adam Lebor, the author of *Tower of Basel*.

In *East Meets West* (chapter 11), the B.I.S. is described as extremely Machiavellian, especially with its connections to fascism and various intelligence circles. In the following section, the bank's historical roots, as well as the role it plays in the global financial system, will be uncovered.

As mentioned previously, the Bank for International Settlements is based in Basel, Switzerland, and was set up in 1930 as a way of enforcing reparations to Allied powers after the Germans lost World War One. But throughout the years, its primary role has changed from one of overseeing wartime reparations, to becoming a powerful financial regulator.

Dubbed the central bank for the central banks, some reports suggest, that it was the only foreign agency of its kind to interact with the German Central Bank (The Reichsbank) during W.W.II. Then in more recent times, it has welcomed the central banks of Russia and China into the fold, giving it unprecedented financial power over the communist countries.

Therefore it must be asked, what makes the B.I.S. so unique? Firstly, it operates under an international treaty like the United Nations, forcing Swiss authorities to obtain permission to enter the property, while not allowing its assets to be seized, or books audited. Also, it is known to have shares that are owned by some of the big central banks, so it appears the private international bankers control it. Furthermore, because of its tiered structure, the Federal Reserve, Bank of England, Bundesbank of Germany, and the European Central Bank (ECB), dominate core decision making, while less influential banks are placed in various layers around the hub.

So what is its purpose today? Primarily, the B.I.S. is the location where many of the world's top central banks conduct regular meetings, for the purpose of streamlining the global financial system. Therefore, its job is to ultimately help bring about the formalisation and eventual unification of the world's currencies. The first step in that plan was the creation of the European Central Bank, and its currency, the Euro. The B.I.S. was able to achieve that goal because it had the power to manipulate individual European economies, and thus align them into a single currency. And since no one can understand how the bankers did it, most people think the Euro came about organically.

According to many researchers, the Euro is the template for other continental currencies in the future, with media outlets in the U.S. having previously alluded to a new North American currency called the Amero.

There can be no underestimating the power the central banking system wields over the world, and that includes financial assistance for some of the biggest political changes seen throughout the twentieth century. Secret money funnelled from the central banks to clandestine wars, revolutions, and coups-d'etat are so well hidden, that most people and government officials have no idea they are paying for them.

So, nothing has changed since the times of Ur (Babylon) and Nippur (Media-Persia). Through their military might, these great kingdoms of the past were able to protect and extend their economic power. That tradition has continued on through various kingdoms and City-States, and eventually into the British Empire during the eighteenth and nineteenth centuries. Then in the twentieth century, it has become the Anglo-American age, with the City of London remaining the heart of world finance, while Washington has been given the mantle of world policeman.

In the following chapters, it will describe how the elite will try to establish their 'One World Socialist Government', and that they intend to introduce it through fair means or foul.

CHAPTER FOURTEEN

WORLD GOVERNMENT, BY CONSENT

"We shall have world government, whether or not we like it. The question is only whether world government will be achieved by consent or by conquest."

–James Warburg, International Banker, 1950

WHEN STUDYING THE globalist plan one startling fact begins to emerge, and that is, there appears to be a double-pronged attack upon the people of the world. Such an attack comes in two distinct forms, firstly, there is the long game, which includes infiltration and indoctrination. Then there is the short game, where wars and revolutions are used to bring about a quick decisive outcome. In this chapter, it describes the long game, and how the *financial elite* has used their money and power to change every area of our lives, including the way we think.

It could be said that there are two forms of free enterprise at work today, the first is the type we are all familiar with, the one where you make a product, or supply a service and charge a fee, hoping to make a profit. The second type is different though, it is a form of business that hates competition, bribes government officials, uses illegal tactics, avoids taxes at any cost, and works with like-minded people in an effort to amalgamate large tracts of the economy. To understand this second type of free enterprise (ie fascism), and how it has grown to dominate the political world, we need to look no further than Cecil John Rhodes. While other powerful individuals and groups make a concerted effort to remain in the shadows, Rhodes, through his brash personality and single-mindedness, gives a glimpse into how a powerful clique of financiers has infiltrated the political arena.

CECIL RHODES

Cecil Rhodes was born on the 5th of July 1853, in Bishop Stortford, England. He was one of thirteen children and suffered greatly from asthma and tuberculosis, so he was sent to the warmer climes of South Africa at the age of seventeen. After his arrival, he and his older brother Herbert grew cotton in the harsh environment of Natal, but that wouldn't last long, because this part of southern Africa was coming down with 'diamond fever'. So before long, the brothers made their way to the town of Kimberly.

In the beginning, Cecil wasn't really interested in diamonds, but that changed when a large diamond field was found on a farm owned by the De Beer's brothers, so the Rhodes boys gave up farming for good, and went in search of the elusive gems.

While miners thought diamonds only existed in certain types of ground, Rhodes carried on digging and soon struck it rich in what is called 'blue ground'. It was that fortuitous discovery which made him a major player in the industry, and along with business partner C.D. Rudd, Rhodes established the De Beers Mining Company in 1880.

When large deposits of various minerals were discovered in southern Africa, British policy towards the region changed from one of indifference to active colonisation. That policy change helped Rhodes expand his empire, and by the 1880's it gave him the confidence to enter politics. It is at this point that Rhodes true imperialistic ambitions were beginning to show, and this was clearly seen when he was trying to negotiate with the Ndebele people.

If he was successful, he would gain access to large mineral deposits inside the vast African interior. But when the Ndebele people discovered that mass colonisation was taking place, conflict became inevitable. That forced Rhodes to obtain a Special Charter from the British throne, thus giving him the troops he needed to crush any uprising. As history tells us, all resistance was extinguished by 1893, and the vast swath of untamed land would eventually become known as Rhodesia.

To understand the mindset of Cecil Rhodes, one ambitious plan he had, was to build a railroad from Cape Town to Cairo. A project that would open Africa's natural resources to development, and thus British Imperial domination, but because of his deteriorating health and eventual death at the age of 48, his dream of a single Africa under British rule would never come to fruition.

ROUND TABLE

There was another side to Cecil John Rhodes, one that wasn't just economic and political expansionist, but one of shadowy manipulator of world affairs. At the age of twenty, while attending Oxford University, Rhodes came under the teachings of John Ruskin, a man who believed Imperial Britain should expand its boundaries and rule the entire world.

It was after their encounter that Rhodes became more than just a British patriot, with his whole existence beginning to form around the ideals of British expansionism. The following statement shows just how far this powerful individual was prepared to go.

"To and for the establishment, promotion and development of a secret society, the true aim and object whereof shall be for the extension of British rule throughout the world, the perfecting of a system of emigration from the United Kingdom, and of colonialization by British subjects of all lands where the means of livelihood are attainable by energy, labour and enterprise, and especially the occupation by British settlers of the entire Continent of Africa, the Holy Land, the valley of the Euphrates, the Islands of Cyprus and Candia, the whole of South America, the Islands of the Pacific not heretofore possessed by Great Britain, the whole of the Malay Archipelago, the seaboard of China and Japan, the ultimate recovery of the United States of America as

an integral part of the British Empire, the inauguration of a system of Colonial representation in the Imperial Parliament which may tend to weld together the disjoined members of the Empire and finally, the foundation of so great a power as to render wars impossible, and promote the best interests of humanity."

–Cecil Rhodes
Confession of Faith,
1877

That quote, among many others, showed the ideology of the man, and because he never married or had children, he could dedicate his life and wealth to achieving those ideals. But because of his deteriorating health, Cecil Rhodes had to write a series of 'wills' to help fulfil his vision for consolidating political power, under the umbrella of elite financial interests. According to Professor Carroll Quigley in his book *Tragedy and Hope*, Rhodes, along with William T. Stead, Bree (Lord Asher) and Lord Alfred Milner, set about organizing the inner-circle of what became known as the Round Table Groups. Even though the diamond magnate died in 1902, it was Milner and his control of the Rhodes Trust that financed the secretive organisation. One statement Professor Quigley made pertains to how quickly the society established itself.

"By 1915 Round Table Groups existed in seven countries, including England... (and) the United States... Since 1925 there have been sub-

> *stantial contributions from wealthy individuals*
> *and from foundations and firms associated with*
> *the international banking fraternity..."*

–Carroll Quigley
Tragedy and Hope,
1966

Professor Quigley went on to say, how many companies like J.P. Morgan, Lazard Brothers, as well as families like Rockefeller and Whitney, used their wealth and influence to help push the organisation into new countries. Also, it is interesting to note, that Milner was an agent for the powerful house of Rothschild, whom it turns out, were also the financiers for De Beers.

When it comes to the Round Table Groups, one of their most coveted objectives is to stay under the radar. To do that, the Royal Institute of International Affairs in London established an outer-ring called The Association of Helpers. This allows ordinary businessmen to join the fraternity, thus giving it an air of inclusiveness. Then for high-ranking initiates there exists an inner-core called The Society of the Elect, and it is this core that promotes the hidden agenda.

Also, many countries operate under different names, for example, in the United Kingdom they are the Royal Institute of International Affairs, while in the USA they go by the name of Council on Foreign Relations. Some say that organizations like the Trilateral Commission and Bilderberg's are just elite branches of the R.T.G's.

When it comes to their main objective, their agenda is to influence the political process in each country for the benefit of the *financial elite*. By grooming inner-core members, those individuals can then be placed inside important government departments, which in turn gives the R.T.G.'s. influence over the government regardless of which party is in power. Again, insider Carroll Quigley flatly states how the elite have no time for opposing political parties.

> *"The argument that the two (political) parties should represent opposed ideals and policies, one, perhaps of the right and the other of the left, is a foolish idea acceptable only to the doctrinaire and academic thinkers... Instead the two parties should be almost identical, so that the American people can 'throw the rascals out' at any election without leading to any profound or extensive shifts in policy."*
>
> **–Carroll Quigley**
> **Tragedy and Hope,**
> **1966**

To give one example, here is a list of US government positions in both the Democratic and Republican parties that CFR members have served on.

C.F.R. MEMBERS IN THE US GOVERNMENT		
Presidents	**Vice Presidents**	**Secretary of State**
Herbert Hoover	Richard M. Nixon	Henry L. Stimson
Dwight D. Eisenhower	Hubert Humphrey	Edward R. Stettinius
John F. Kennedy	Gerald R. Ford	Dean G. Acheson

Richard M. Nixon	Nelson A. Rockefeller	John Foster Dulles
Gerald R. Ford	Walter Mondale	Christian A. Herter
James E. Carter	George H.W. Bush	Dean Rusk
George H.W. Bush	Richard Cheney	William P. Rogers
William J. Clinton		Henry A. Kissinger
George W. Bush		Cyrus R. Vance
Barack H. Obama		Edward S. Muskie
		Alexander M. Haig
		George P. Shultz
		Lawrence Eagleburger
		Warren Christopher
		Madeleine Albright
		Colin L. Powell
		Condoleezza Rice

Secretary of Defence	Secretary of the Treasury	C.I.A. Director
Henry L. Stimson	Andrew W. Mellon	Walter Bedell Smith
Robert P. Patterson	Ogden L. Mills	Allen W. Dulles
James B. Forrestal	William H. Woodin	John A. McCone
Robert A. Lovett	Henry Morgenthau Jr	Richard Helms
Neil H. McElroy	Robert B. Anderson	James Schlesinger
Thomas S. Gates Jr	C. Douglas Dillon	William E. Colby
Robert S. McNamara	Henry H. Fowler	George H. W. Bush
Melvin R. Laird	David M. Kennedy	Stansfield Turner
Elliot L. Richardson	George P. Shultz	William J. Casey
James Schlesinger	William E. Simon	William H. Webster
Donald H. Rumsfeld	W. Michael Blumenthal	Robert M. Gates
Harold Brown	G. William Miller	R. James Woolsey
Casper W. Weinberger	Donald T. Regan	John M. Deutch
Frank C. Carlucci	Nicholas F. Brady	George J. Tenet

Richard B. Cheney	Lloyd Bentsen	Michael Hayden
Les Aspin	Robert Rubin	
William Perry	Lawrence W, Summers	
William S. Cohen	Henry M. Paulson Jr	
Robert M. Gates		

Through the London based Royal Institute of International Affairs, along with its many off-shoots like the Council on Foreign Relations, the elite has been able to influence the way politics has moved throughout much of the twentieth century.

Additionally, they have also been able to use their global off-shoots to subvert the sovereign laws and business practices of nations. To do that, the R.T.G's. use global issues, or crisis, so they can introduce the solution for governments to follow. According to its own website, the CFR uses Think Tank's to help create closer cooperation between nation-states.

> *"The Council on Foreign Relations has launched an international initiative* [Think Tank] *to connect leading foreign policy institutes from around the world in a common conversation on issues of global governance and multilateral cooperation* [global agenda]. *The mission of the 'Council of Councils' is to find common ground* [influence sovereign nations] *on shared threats, build support for innovative ideas, and inject remedies* [problem solve] *into the public debate* [produce media propaganda] *and policy making processes of member countries* [influence govern-

*ments]... C.F.R. will convene the inaugural
'Council of Councils' conference on March
12-13 in Washington DC. Participants will
tackle four major themes at the first gathering:*

- *the overall state of global governance and
  multilateral cooperation.*
- *the status of the nuclear non-proliferation
  regime (with a focus on Iran).*
- *the dollar's future as the world's reserve
  currency.*
- *The criteria for humanitarian intervention, in the wake of regime change in
  Libya and the ongoing crisis in Syria. "*

**–Council on Foreign Relations,
9th March, 2012**

The above statement is one example of how the various
Round Table Groups work together to coerce the world's
nations into following a common global agenda. The
non-italicized inserts placed in the quote are to show
what they are truly up to, and that they work their agenda
from the top down, as well as from the bottom up. With
the elite, nothing is left to chance, and whether it is true
events or fabricated lies, they will use anything they can
to establish the foundations for a single economic, political, and social order throughout the world.

In his book *None Dare Call It Conspiracy*, Gary Allen
states how some members of the CFR are also members
of organisations that want the wholesale abolition of
western democracy.

"Also in the C.F.R., are men from such openly Leftist organisations as the Fabian Socialist Americans for Democratic Action, the avowedly Socialist League for Industrial Democracy—(formerly the Intercollegiate Socialist Society), and the United World Federalists which openly advocates world government with the Communists. Such devotedly Socialist labour leaders as the late Walter Reuther, David Dubinsky, and Jay Lovestone have also been members of the C.F.R."

–Gary Allen
None Dare Call It Conspiracy,
1971

Members of the R.T.G's. wear many hats, for many occasions, and they have no trouble financing unsavoury groups if it suits their purposes. For their ultimate objective is to use any person or organisation to get what they want, regardless of how radical the group is.

For the elite like Milner, Stead, Asher, and Rhodes, no cost is too great, especially if it means more wealth and power for them and their co-conspirators.

Speaking of Rhodes, it may have been his intention to set up the Round Table Groups in the hope of expanding British Imperial Rule, but it appears that his endeavours failed, for we now know the British Empire fell from its lofty position in the decades following his death. On the other hand, the Round Table Groups have extended the influence of the City of London, like most people could never have imagined.

And the irony is, Rhodes may have thought he was in the inner-clique but in the end, may have been nothing more than a pawn in a game much larger than him.

When it comes to managing Foundations like the Rhodes Trust, secrecy is the objective, and if it hadn't been for one report in the 1950's, we still wouldn't know how far their influence extends, and why they are so dangerous to our way of life.

THE REECE COMMITTEE REPORT

"Some even believe we are a part of a secret cabal working against the best interests of the United States, characterizing my family and me as 'internationalists' and of conspiring with others around the world to build a more integrated global political and economic structure — one world, if you will. If that's the charge, I stand guilty and I am proud of it."

–David Rockefeller
Memoirs,
2003

When we think of Foundations or Endowment's, we usually think of charitable philanthropy by well-intentioned people, but that is not always the case. There is a group of Foundations with unimaginable wealth and power, and their sole objective is to undermine our democratic rights by collectivising all nations into a single socialist world order.

The story of these powerful institutions begins in the early twentieth century, when the American people were becoming weary of the robber barons, and their corporate cartels like Standard Oil and AT&T. From that mistrust, a series of anti-trust cases and government reports into how they operate began to appear. One such report between 1952 and 1954, uncovered what appeared to be Un-American activities at the Rockefeller and Ford Foundations, as well as the Carnegie Endowment.

Named after Congressman B. Carroll Reece (R-TN), and led by attorney Norman Dodd, the 'Reece Committee Report' led to one of the most in-depth investigations ever conducted into the giant tax-exempt Foundations. The report was so damning of the *financial elite* and their Un-American activities, that all but three of the reports were either bought up or confiscated, leaving most Americans, and people of the West, without any clue as to how their lives are being manipulated.

In an interview with author G. Edward Griffin in the early 1980s, Norman Dodd revealed some of the finer details from his time on the 'Reece Committee'; here are some of those details.

In one interview Roan Gather, the president of the Ford Foundation, made it clear to Dodd that the primary aim of the Foundation was to use its vast grant-making power to alter life in the United States, so it could be merged with the Soviet Union. That could be one reason why the western media doesn't present the same amount of negative press concerning communist leaders like Joseph Stalin, Fidel Castro and Mao Zedong, considering their atrocities were as bad, if not worse than Adolf Hitler's.

After studying how best to bring socialism to the West, the Foundations quickly came to the conclusion that there were two main avenues by which to bring their dream about. The first was to use the art of war for an effective short-term change, while the long-term approach would involve indoctrination through the media and education system.

In an interview with Dodd, Joseph Johnson, the president of the Carnegie Endowment, reveals how serious the foundations were when using war for their own ends. According to Dodd, the board members including Joseph Johnson were new to their positions, so were naïve in how to treat the investigators. When Dodd asked Johnson, for information about the organization, Johnson only allowed one investigator two weeks to go over their records in New York. He knew it would be a hopeless task for one person to look over fifty years of records. But Dodd was up to the task, he sent in his young colleague Kathryn Casey with Dictaphone belts, and told her to only go over certain dates and leave the rest. From those investigations, the shocking truth began to emerge about how best to get the United States into a war, or series of wars, so the elite could bring about rapid change to the country.

In 1908, shortly after the creation of the Carnegie Endowment, the insiders began planning the takeover of the US State Department. To achieve this, they would place their own people into the top positions. This was important because the State Department is the middle-man between the US government and all other foreign powers. With their control of the media, and now the State Department, the Foundations could set about getting the American people into WWI.

For the elite, the introduction of communism and a

world governing body were two of the most important issues on their agenda. And as history tells us, Russian communism was established during WWI, and made great strides during WWII. As for the introduction of a world governing body, it was reluctant US president Woodrow Wilson, who put paid to the League of Nations when he refused to sign it into law. But that failure would be rectified after WWII, with the introduction of the United Nations. One interesting footnote about the United Nations is that when the building was constructed in New York city, the land was donated to it by none other than the Rockefellers.

Also taken from the minutes of the Carnegie Endowment, the second avenue the foundations used, was the long-term approach of re-education. For that task, the Carnegie Endowment would team up with the Rockefeller Foundation, for the express purpose of rewriting the school textbooks. After many years of effort, the foundations were able to introduce socialist and humanist doctrine into the schools. This is important because it determines what and how our youth are taught, even moulding their moral character of what is right and wrong. The list below shows just how far our western schooling system has changed over the last one hundred years, and what has now become acceptable.

Unacceptable	Acceptable
Free Enterprise	Corporate takeover
Patriotism	Globalization
Morality	Tolerance
Private Property Rights	Collective law

Self Defence	Police State
Rights of the Parents	Rights of the State
Self-Education	State Education
Religious Freedom	Political Correctness
Creation	Evolution
Democratic Freedom	Socialism
Freedom of Speech	Hate Speech

In the six decades since the 'Reece Committee Report' was released, the western education system has well and truly been taken over by the socialists. The vast wealth of the giant tax-exempt foundations has allowed the elite to implement their plans in the West, while most of the general public had no idea what was going on. In the end, the 'Reece Committee' had to be closed down prematurely because of incessant opposition from within the US government. From the report, B. Carroll Reece came to two obvious conclusions. The first was, there was definitely a conspiracy to overthrow the United States, and secondly, the US tax-payer was helping to pay the bill.

It's clear the long game (infiltration and indoctrination) is a powerful tool for subverting the natural progression of any country, and whether it is used over decades or centuries, it usually leaves its mark upon the generations that follow. But there is another approach to changing a country, and because it can create massive upheaval, it usually leaves many innocent people homeless, or even worse, dead.

CHAPTER FIFTEEN

WORLD GOVERNMENT, BY CONQUEST

"In the councils of government, we must guard against the acquisition of unwarranted influence, whether sought or unsought, by the military-industrial complex. The potential for the disastrous rise of misplaced power exists and will persist."

–US President, Dwight D. Eisenhower

THROUGH THE IMPLEMENTATION of the short game, the elite have perfected the age-old art of subversion, so they know how to bring rapid change to a country, whether it is through financial manipulation or clandestine warfare. Those changes are not usually possible by any other means, and almost always lead to devastating consequences for the people affected.

Throughout history, there has always been a link between money and the military, but in more modern times, this connection is seen nowhere more clearly than with the British East India Company, or what some affectionately call The Company.

With its headquarters in the City of London, the British East India Company was formed on the last day of 1699, as a counter to the economic might of the Dutch East India Company. It was comprised of wealthy businessmen, who banded together to form a powerful trading company that imported products like tea, silk and spices from various countries around the world. In the subsequent years, the East India Company, or EIC, would become one of the most powerful trading companies in the world, leading some to suggest that it was like an empire within an empire, making its own independent agreements like it was a sovereign state.

What made the EIC so unique was its use of a private army in India, and by 1803, that military force had grown to 260,000 Indian soldiers, all the while retaining the full backing of the British Crown. An investigative article in *The Guardian* news site provides a good description of just how ruthless the relationship between the State and the EIC had become.

> *"100 years into its history, it had only 35 permanent employees in its head office. Nevertheless, that skeleton staff executed a corporate coup unparalleled in history: the military conquest, subjugation and plunder of vast tracts of Southern Asia. It almost certainly remains the supreme act of corporate violence in world history. For all the power wielded today by the world's largest corporations... they are tame beasts compared with the ravaging territorial appetites of the militarized 'East India Company'... When it suited, the E.I.C. made*

much of the legal separation from the government. It argued forcefully, and successfully, that the document signed by Shah Alam—known as the Diwani—was the legal property of the company, not the crown, even though the government had spent a massive sum on naval and military operations protecting the E.I.C.'s Indian acquisitions. But the M.P.s who voted to uphold this legal distinction were not exactly neutral: nearly a quarter of them held company stock, which would have plummeted in value had the Crown taken over."

**–The Guardian,
4 March, 2015**

Ruling the seven seas during the 18th and 19th centuries, the British Navy was able to open trade routes with nations throughout the world, and from that trade, open the doors to British immigrants, as well as their culture. So it must be remembered that without the Royal Navy, their expansionist programs would not have been possible.

Despite its ruthless conduct, the British East India Company lost much of its power after the Great Indian Uprising of 1857. Then under Queen Victoria, the Crown had to nationalize the company, but even that couldn't stop its inevitable demise in 1874.

While the EIC has faded into history, the corporate-government relationship by which it operated has not. If anything, this form of governance has become more entrenched with every decade, even outgrowing the confines of the British Commonwealth, by establishing its monopolistic philosophy around the world.

In the twentieth century, the relationship between the military and the *financial elite* has grown, with western corporations expanding their wealth and influence through military protectionism. Even though many people are familiar with corporate imperialism, there appears to be another darker side that is not widely understood. It comes in the form of dualism. What that means is, some of those same companies who supply western countries with goods and services, also supply most of our enemies. And some of those goods just happen to include advanced weapons systems.

So that raises an interesting question, why would these pillars of free enterprise support regimes that appear anti-democratic, and against everything they supposedly stand for? As previously mentioned in *East Meets West* (chapter 11), the answer is, the *financial elite* help finance and maintain all political systems, so they can control the ultimate outcome.

Based around the research of Professor Antony C. Sutton, while working at the Hoover Institute, the next two sections of this chapter describe the financial and industrial help various socialist/fascist systems received from the elite.

ASSISTED NAZI SOCIALISM

The relationship the *financial elite* had with anti-democratic forces is seen nowhere more clearly than in Germany before and during World War Two. Historians tell us that millions of allied troops died trying to free the world from German aggression, and had it not been for our superior number of tanks, aeroplanes and ships, the allies would not have had the ability to get the job done. What our allied troops didn't know, was that those same companies who were producing our armaments, were also producing vital equipment and supplies for the German war machine.

After WWI, Germany was financially crippled, so how was it possible for them to contravene the Versailles Accord (heavily restricted military) and build a technically advanced military within twenty years, and ten of those years during the Great Depression?

It is believed that without outside assistance, and in particular American corporations, the Nazi military-industrial machine could never have got off the ground. Through subsidiaries and co-operative understandings, scientific and technological transfers were put in place so the Germans could advance their industry. But not just that, financial assistance was also given so they could produce the necessary technologies en masse.

When it comes to Wall Street and its connection to fascist Germany, you cannot go past the giant chemical company I.G. Farben. Leading up to the war, the director of I.G. Farben-Germany just happened to be Max Warburg, while his brother Paul, was the director of I.G. Farben-America. This is the famous banking family who helped create the privately run Federal Reserve. But it goes much deeper than flagrant nepotism, in fact, I.G. Farben supplied about half of Hitler's funds for the National Socialist party when it took power in 1933. In his book *Wall Street and the Rise of Hitler*, Antony Sutton describes in more detail how the Wall Street-I.G. Farben connection worked.

"In brief, 45 percent of the funds for the 1933

election came from I.G. Farben. If we look at the directors of American I.G. Farben—the U.S. subsidiary of I.G. Farben—we get close to the roots of Wall Street involvement with Hitler. The board of American I.G. Farben at this time contained some of the most prestigious names among American industrialists: Edsel B. Ford of the Ford Motor Company, C.E. Mitchell of the Federal Reserve Bank of New York, the Standard Oil Company of New Jersey, and President Franklin D. Roosevelt's Georgia Warm Springs Foundation."

–Antony C. Sutton
Wall Street and the Rise of Hitler,
2002

Even though this happened before the outbreak of World War II, it uncovers the Wall Street connection, and the influence it had over the future direction of Germany. Antony Sutton goes on to say how three board members of American I.G. Farben were found guilty at Nuremberg, yet according to the records, the men and companies mentioned in the above quote, were not even questioned about the Hitler Fund. This just goes to show that Wall Street was present when Hitler rose to power, then when the chips were down at the end of the war, the American industrialists just turned on their heels and melted quickly into the background.

Another important connection I.G. Farben had, was with Standard Oil (Rockefeller) of New Jersey. The German chemical cartel, who just happened to be the manufacturer of Zyklon B for the death camps, required the secrets to manufacturing synthetic fuel. The reason for that was simple, Germany lacked oil for its war machine, but the one thing it did have in abundance was coal. So what did they do, Standard Oil, through a co-operative agreement with I.G. Farben, gave the Germans the secret to synthetic oil, thus allowing the Nazis to convert coal into petroleum products on a large scale. The transfer was so brazen that when US government officials denied Standard Oil permission to send the technology, the directors sent it anyway. It must be remembered that, without this important piece of technology, the Nazis might not have been able to go to war in 1939.

In any modern war, military vehicles are a necessity, and while allied soldiers were fighting and dying in war machines built by corporations like Ford and General Motors, those same two companies were also building machines for the Nazis. Ford built vehicles for the axis powers, from its German plant in Cologne, as well as its French, Belgium and Dutch subsidiaries while under German occupation. This gave the Germans a vital supply of trucks as well as aircraft engines for the Luftwaffe.

It must be asked though, were the Ford plants willing participants, or were they simply nationalised for the German war effort? Again, Antony Sutton describes how the Foreign Funds Control Section of the US Treasury Department initiated an investigation into some of the big US firms operating in Nazi-occupied Europe. Here is a portion of the report that involved the French division of Ford and its US parent between 1940 and 1942.

"(1) the business of the Ford subsidiaries in France substantially increased; (2) their production was solely for the benefit of the Germans and the countries under its occupation; (3) the Germans have "shown clearly their wish to protect the Ford interests" because of the attitude of strict neutrality maintained by Henry Ford and the late Edsel Ford; and (4) the increased activity of the French Ford subsidiaries on behalf of the Germans received the commendation of the Ford family in America."

–F.F.C.S. Report

The board of directors at Ford-USA knew exactly what was happening with its foreign division in France, and must have been in full agreement with how the Germans were using it in their war effort.

When it comes to the money trail, it isn't known how much of the profits produced at Ford's European factories left there for the parent company during the war. But according to Ken Silverstein in an article in *The Nation*, Ford-Germany (known as Ford-Werke) set aside dividend payments to Ford-USA during that period. In their defence, Ford claims it only received $60,000 in dividends, but again, Silverstein states that Ford-USA was a private company up until 1956, and they refuse to release any financial records from the war period.

In addition, Ford-USA was probably fully aware of what was happening at its other divisions in Nazi-occupied Europe. For example, Maurice Dollfus of Ford-

France, sent a letter to Edsel Ford in 1941, explaining how delighted he was at receiving 1.6 million francs in business from the Nazis, which included 20 trucks a day for the Wehrmacht. In reply, Edsel Ford said he was happy to hear they were making progress, considering the conditions they were working under.

Another financial tie-in involved the RAF bombing of the Ford plant in Poissy France during 1942. While pictures of the bombing appeared in American papers, the articles never mentioned who the plant belonged too. Fortunately for the automaker, the pro-Nazi, Vichy government, paid Ford 38 million francs in compensation for damage inflicted at the plant.

Lastly, when it comes to the question of neutrality versus collusion, we need to see whether Ford was a willing participant in helping the Nazi regime gain any type of military advantage. According to a letter sent from Heinrich Albert, a Ford lawyer in Germany, to Ford-USA executive Charles Sorenson, it appears the latter might be the case. In the letter he states how happy he was that Ford had denied requests by Roosevelt and Churchill to increase military vehicle production for the allies. Even though Ford's stance in the US and UK didn't last long, it gave Ford-Germany the advantage of ramping up production early on.

There were many big US firms operating in Germany at that time, but the Ford case is important because it clearly shows, how certain US firms were fully aware of what was going on during the war, and for a small portion of the directors at least, there had to be some kind of ideological affinity.

Another auto company operating in Nazi Germany

was General Motors. From its Opel division, the company produced tanks, trucks and armoured vehicles before and during the war. Also, companies like AEG (General Electric Germany) were major contributors to the National Socialist movement before the war, and became the main supplier of electrical equipment for the Nazis when fighting broke out.

Other companies included ITT (International Telephone and Telegraph), who supplied important technical assistance and also held significant investments in Focke-Wolfe, the military aircraft manufacturer. Likewise, support came from companies like Texaco Oil, Ethyl Corporation, and computer giant IBM (International Business Machine), whose punch card system helped facilitate Nazi genocide.

While some might suggest that this was just simple profiteering by faceless money hungry companies, again the truth appears to be quite different. According to Antony Sutton, those big companies that operated in Germany in the 1930's could not take their profits out of the country, so whatever they invested wouldn't have benefitted the parent company. Furthermore, there is evidence suggesting companies like Ford-USA weren't just neutral but were in collusion with the Nazis, and may have collected substantial profits from German-occupied Europe during the war.

As mentioned before in this book, it appears the *financial elite* have the same ideology as the National Socialists (Nazis), and that the German fascists were nothing more than pawns in a game much larger than themselves.

ASSISTED SOVIET SOCIALISM

"Stalin paid tribute to the assistance rendered by the United States to Soviet industry before and during the war."

—Extract from a report by Ambassador Averell Harriman in Moscow, 30th June, 1944

The dream of Russian socialism (communism), was to establish an environment where there was no need for private property or free enterprise, making self-determination from an individual perspective unnecessary. It might sound achievable in theory, but in practice, it goes against all that is inherent in man's character; even Lenin himself realized that pure communism didn't work. That's why, just prior to his death in 1923, he began a program of economic reforms, which opened the way for large western corporations to trade with the Soviet government.

From its inception, the Soviet Union and its allies (including China) have been recipients of vast quantities of finance, food and technical aid, and without these secretive arrangements, communism would have died a natural death decades ago.

Among many treasonous acts committed by western companies, one prime example was when the Ford Motor Company, was given the contract to build and equip the vast Gorki automobile plant at Nizhniy-Novgorod (formerly Gorki). Later on, the Gleason Company of

Rochester, New York, equipped the plant with machinery that allowed the Soviets to produce military type trucks on a mass scale. A quote from Professor Antony C. Sutton's book *The Best Enemy Money Can Buy* shows the repercussions this treasonous relationship spawned.

> *"Many of the trucks used on the Ho Chi Min trail were GAZ vehicles from Gorki. The rocket launchers used against Israel are mounted on GAZ-69 chassis made at Gorki. They have Ford-type engines made at Gorki."*
>
> **–Antony C. Sutton**
> ***The Best Enemy Money Can Buy***

From the relatively small investment the Soviets paid Ford to build the plant at Gorki, the return has been exponential. To start with, it took Ford many years, and millions of dollars, to perfect the assembly line mass production process. In addition, Ford and other companies, handed over designs for vehicles that the Soviets and their allies could never have developed on their own. It's this kind of assistance from western companies that allowed the communists to become an equal in military hardware.

Ford wasn't the only company committing treason. In fact, when it comes to sensitive technological transfers, it's a veritable who's–who of western companies clamouring to get behind the Iron Curtain. Take, for instance, the Togliatti vehicle plant. It was bigger than the Gorki plant, and built under contract by Fiat S.p.A. in 1966, with US firms sub-contracted to supply machine parts. The plant mass-produced the Lada, which was a direct copy of the Fiat-124, but it was also equipped to produce military equipment, like engines and parts for their military Jeep equivalent, a vehicle used in many of the world's war zones.

Then in 1976, the gigantic Kama River truck plant was opened, and it was capable of producing 100,000 large vehicles for both civilian and military applications. According to Sutton, machinery and parts were supplied by firms like Gulf and Western Industries, Honeywell and Swindell-Dresser Company (a subsidiary of Pullman). And adding insult to injury, these plants were built using US taxpayer-backed loans through the Chase Manhattan Bank, and all this while the West had been fighting a war with the Soviet-backed communists in Vietnam.

Western-backed funding processes were used many times for different projects behind the Iron Curtain. One project of note was the steel mill and fabrication plant at Magnitogorsk. When it was built in the 1930's, it was the biggest plant of its type in the world.

As for Soviet aircraft, Sutton states that western companies were transferring sensitive equipment and knowledge for many years, giving the Reds a formidable combat airwing. Going back to the 1930's, companies like Douglas Aircraft, handed over full schematics and assembly processes for the DC-3, a plane the Soviets later released as the PS-84. Other companies like Birdsboro Steel Foundry and Machine Company sent in giant presses critical for making plane parts, while The United Engineering and Foundry Company supplied advanced material technologies, the list goes on and on.

After the war, captured German scientists helped get

Soviet jet engine technology off the ground, but it wasn't until they bought fifty-five super advanced centrifugal turbo-jets from Rolls Royce, that they were able to leap forward and match anything the West could produce. This was proven when the Soviet-backed North Koreans used their MIG-15s to deadly effect against American fighters at the start of the Korean conflict.

Even the similarities in looks and performance, as well as the uncanny release dates for Soviet planes, points towards some kind of secretive arrangement between them and the capitalists.

Western technologies weren't just finding their way into the Soviet air force but were also turning up in naval ship design. According to Antony Sutton, 95% of all ship and engine designs were sourced from western companies like Tosi (Italian engines), Gibbs and Cox (American destroyer designs), and Burmeister & Wain (large diesel engines) to name a few. One interesting case regarding the transfer of advanced naval technologies, involved the latest Soviet submarines.

No one is denying the KGB hasn't used espionage to acquire sensitive information for its formidable submarine fleet, but they could not have kept up with the West if it hadn't been for secretive corporate arrangements. One instance of collusion was when the giant Japanese corporation Toshiba, was caught selling sensitive military equipment in 1983/84. Under the plan, the communists intended to pay Toshiba $30 million to supply advanced lathes for machining giant blocks of brass into non-cavitating props for their latest submarines. Had it not been for media exposure, the public would have been none the wiser to these treasonous acts.

One last example professor Sutton details was when the Soviets were able to acquire MIRV technology for their missiles. Before the 1970s, the Soviets did not have MIRV (Multiple Independently Targetable Re-entry Vehicles) capability, which allowed one nuclear missile to send multiple warheads to several targets at once.

What the Soviets lacked were the high precision miniature ball bearings and races needed to control the MIRV system. Only one company could produce those precision parts, and that was Bryant Chucking Grinder of the USA. After BCG was given permission to send the manufacturing equipment, the Soviets were able to make a giant leap forward in their nuclear strike capability.

Even right up to the very end of what is known as the Soviet Era, western firms like IBM and General Electric were supplying computer technology, so the Soviet weaponeers could enter the information age.

It doesn't just stop with the Russians though; the Chinese have also been recipients of sensitive foreign technology, and if that is the case, then their rise in economic power over the last thirty years has been anything but luck. There again, large western companies along with the Clinton Administration, went in and signed cooperative agreements with the communist government so technology could be transferred en mass. That transfer didn't just involve cars, high speed-trains and construction techniques, but included the latest military equipment, like jet fighters, hardened chips, missiles, radar systems, and even nuclear weapons technology.

It seems that over time the recipients may change, but it's business as usual for the biggest industrial companies on earth, and their masters, the international bankers.

INVISIBLE WARFARE

"I spent 33 years in the Marines, most of my time being a high-class muscle man for big business, for Wall Street and the bankers. In short I was a racketeer for Capitalism."

–U.S. Brigadier General, Smedley D. Butler

Just after the second world war, the relationship between the *financial elite* and the military shifted to a new level, when the United States Defence Department and CIA was created in 1947. These two powerful entities and their estimated 1,000 permanent bases (which are considered US territory) around the globe, give those who control them the power to divert the policy of almost any country they desire.

In the West, we have been taught that the Pentagon and CIA defend the US and her allies from the evils of communism and despots the world over. But that is not the case. When we look at most of the wars, revolutions and coups-d'etat of the twentieth century, it doesn't take long to see who wins when it comes to expanding their wealth and power. There's one group that always seems to come out on top, and when their power is threatened they pull no punches, regardless of who the recipient might be.

There are many ways a government can be overthrown; some are by direct military intervention like the Iraq wars of the 1990's and 2000's. Wars like that make arms manufacturers hundreds of billions of dollars in weapons sales and gives the Pentagon the excuse to stay in the Middle East indefinitely. Then there is another type of war, this one is the sole responsibility of organizations like the CIA, MI6, FSB, and involves the dethroning of leaders and parties by subversive means. It usually happens when countries refuse to play by the rules set out by the *financial elite.*

An example of a bloodless coup can be seen when the Australian people voted Gough Whitlam's Labour party into power in the early 1970's. Upon his re-election in 1975, he stated to the Australian people that he would review the presence of American and British spy bases on Australian soil. When the American and British governments became aware of his plan, they called in the Governor General to dissolve the Whitlam government and place the opposition into power. This sort of war might be bloodless but it is undemocratic and goes against the wishes of the people.

A different type of war, one that isn't bloodless, has proven to be extremely effective for the elite. It is the shadowy indirect wars financed by agencies like the CIA with the help of military advisors and Special Forces from the Pentagon. While acknowledged by the majority of western observers as 'not good', they are seen as a necessary evil for protecting our liberties. However, this could not be further from the truth. There are winners and losers in a bloody coup, and usually, freedom, independence and the people are the losers, while despots and the corporate elite are the winners.

Many bloody coups have happened in recent history, but three accounts mentioned here stand out as blatant greed by the *financial elite*, and show that they will stop at nothing to consolidate their power over the nations of the world. One instance in Guatemala, involved the wholesale confiscation of peasant land, by the giant food corporation United Fruit. Early on, when United Fruit of Boston first moved into the Central American country to set up business, it used greedy local businessmen to buy up land cheaply off the peasants. They would then resell it to United Fruit, making the company the biggest landowner in the country. Then in 1952 Jacobo Arbenz was elected President of the nation, and promised to return large tracts of unused land owned by United Fruit back to the peasants, so they could earn their own living. When the company got wind of this, its backers, which included the Rockefeller family, used the Council on Foreign Relations (CFR), to see how best to handle this threat. By using their influence over the CFR, they were able to persuade the US government with the lie that Arbenz and the peasants were nothing more than communists. It has to be said, no true communist would want to hand the land back to the people, but that didn't matter—the CIA had the mandate to go in and attempt to overthrow the Guatemalan president.

It didn't all go smoothly though. After various failed attempts, the CIA sent planes in to bomb the capital, Guatemala City, and by 1954, Arbenz had been deposed and a CIA puppet by the name of Colonel Carlos Castello-Amas, had been installed. He quickly overturned all the reformist agenda of the Arbenz government and led Guatemala into a thirty year series of military dictatorships which killed an estimated one hundred thousand Guatemalan citizens.

Another coup sponsored by the CIA and MI6, and linked to giant corporations was the overthrow of Mohammad Mosaddegh. The democratically elected Prime Minister of Iran (1951–1953) was ousted by the two intelligence agencies for attempting to nationalize the Iranian oil fields.

Since there discovery in 1908, they had been under the full control of Anglo-Iranian Oil, a company which later became known as British Petroleum (BP). When Mosaddegh was ousted the Shah was installed, and from that time on it was business as usual.

One other important example of corporate-intelligence cooperation was when Salvador Allende was elected president of the Republic of Chile in 1970. As part of the reforms he was trying to introduce, he wanted to repatriate certain assets held by large US corporations, and in particular, the vast copper deposits held by Anaconda and Kennecott corporations. That decision bought the ire of the elite, and the first thing the US corporates did was strangle the Chilean economy by restricting loans, investments and even parts for machinery. It was a cunning move and made it easier to implement a coup-d'etat.

By 1973, the Allende government was overthrown by a CIA-backed military coup, which installed Augusto Pinochet (1973–1990) as the leader. That allowed the Trans-National Corporations to keep control of the Chilean economy, while the dictator had hundreds of thousands of his own people tortured, and tens of thousands killed. Pinochet would eventually be honoured by former British Prime Minister Margaret Thatcher for bringing 'democracy' to Chile, even seeing out his days in the UK, under what they termed "house arrest", or what some might call government protection.

PAID TERRORISM

At the end of the cold war in 1991, it was thought by most observers, that world military and intelligence spending would plummet because of our new détente. So, for the elite, it was necessary to create a new enemy, one that couldn't be easily subdued and would require a global military effort to manage. That new bogey-man, the one that has struck fear into the hearts of many, is International Terrorism. And because of its global reach, it has created giant military and intelligence budgets throughout the world, and cost the lives of many innocent people.

But could there be more to modern terrorism than what we are led to believe? It appears there could indeed be a thread of evidence linking global terror organisations to the *financial elite*, that's because, vast sums of cash have disappeared into shady organizations with links to extremism. And even though some observers in the US government have stated how the missing money is nothing more than bad accountancy, recent news reports have surfaced claiming otherwise. One investigative report from *CNBC* claims how billions of dollars in so-called reconstruction money for Iraq, had mysteriously gone missing, and that it was extremely difficult to uncover its whereabouts.

> *"The New York Fed is refusing to tell investigators how many billions of dollars it shipped to Iraq during the early days of the U.S. invasion... It was one of the largest shipments of cash in history, and the Inspector General says that if the money was stolen, that would represent the largest heist in history ..."*

> **–CNBC,**
> **21st June, 2011**

That particular incident involved $2.4 billion dollars. But overall estimates suggest, well over $100 billion dollars of aid money for places like Iraq and Afghanistan have gone missing in the last fifteen years alone. So where has it all gone? In a news report from the *Wall Street Journal* it was discovered that some of the money was going to banks with links to global terrorism.

> *"The Federal Reserve and Treasury Department temporarily shut off the flow of billions of dollars to Iraq's central bank this summer as concerns mounted that the currency was ending up at Iranian banks and possibly being funnelled to 'Islamic State' militants, according to U.S., and Iraqi officials and other people familiar with the matter... The Iraqi officials believe the money has definitely gone to 'Islamic State' through these auctions... 'Islamic State' in 2014 stole about $100 million from a central bank of Iraq-run vault in Mosul, said a person familiar with the theft."*
>
> **–*The Wall Street Journal*,**
> **3rd November, 2015**

Tens of billions of dollars in cash flowed from the US Federal Reserve to the Central Bank of Iraq, where it was auctioned off to various banks. Included in the auctions were banks traced to Iran, as well as areas with strong terrorist influences such as Mosul (Isis/Isil) and Lebanon (Hezbollah). The article goes on to say that US officials in the Fed and Treasury Department were in the dark about where the money had gone.

Official excuses for bad accountancy would be believable if it had only happened once or twice, but it hasn't. It appears that over the last twenty years, there has been a deliberate transfer of aid money from the Federal Reserve, to groups who profess to be the enemies of our freedom.

While terrorism is the current threat, the rise and fall of other socialist/fascist systems throughout the twentieth century has brought massive changes to the global political landscape. And because the *financial elite* have supplied much of the money and weapons for many of those wars, revolutions and coups d'etat, it's important to understand why these orchestrated events are so important.

Known as Dialectics, it is a system used by the elite for bringing change to nation-states, but because it is done in a certain way, it leaves the populous totally unaware that they have just been played.

HEGELIAN DIALECTIC

Some believe it has been around since the time of ancient Greece and India, while others say the dialectical method is much older than that. Its origins may be uncertain, but its implementation over the last couple of centuries has changed the world so profoundly that the political and economic landscape of today would look totally unrecognizable to the people of centuries past.

In essence, the dialectical method is an ideology based on the premise, that if two or more people hold different points of view on a common subject, they can come to a single truth through reasoned arguments. While various forms of this ideology have been around for a long time, it was a philosopher by the name of George Wilhelm Friedrich Hegel (1770—1831) who took the dialectical philosophy to another level. His basic proposal was, if you want to bring change to an idea or system, present its exact opposite in an acceptable way, and the original idea or system will change in some way. This has become known as the Hegelian dialectic. Hegel's simple formula was:

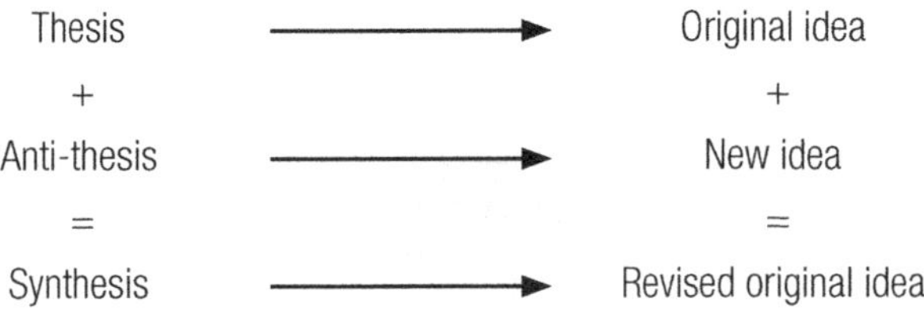

This ideology of Hegel's may have fallen by the wayside like many others, had it not been for two men named Karl Marx and Frederick Engels. In 1847, these two were part of London's Communist League, and their job was to take ancient feudalistic ideas and rework them into the Communist Manifesto. Their plan was to use a more advanced version of Hegel's dialectics to quietly introduce their manifesto upon an unsuspecting populous.

So, what does this have to do with the *financial elite*? What this means is, two opposing political systems (democracy and communism) can be used to destroy each other, so a new political system (world government) can emerge in their place. Using this concept in the real world, a series of wars, and economic cycles can be engineered to bring mankind willingly into a final solution, and all without the knowledge or consent of the world's people.

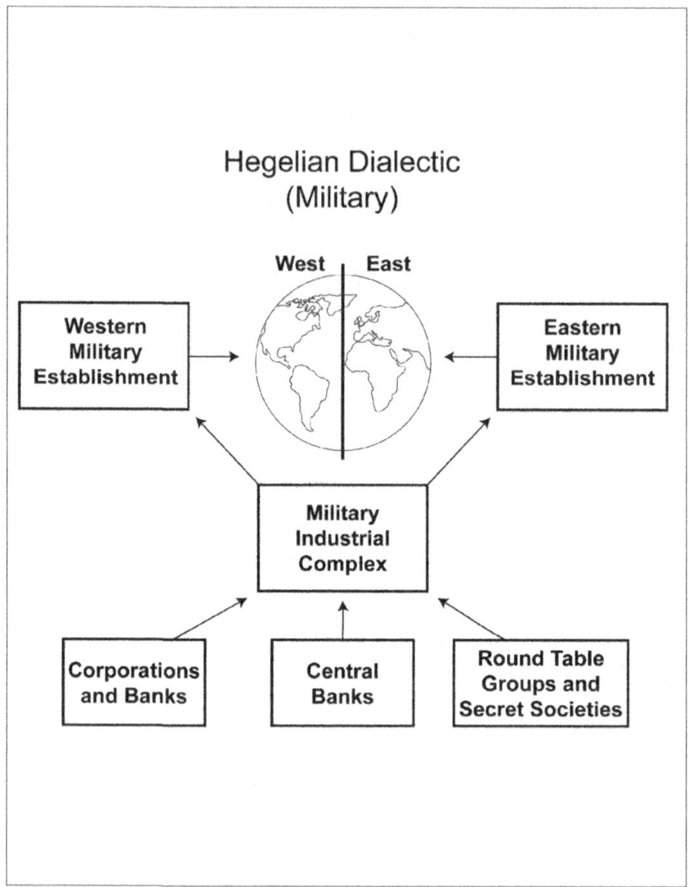

Hegelian Dialectic (Military)

By controlling both the eastern and western military establishments, the *financial elite* are able to create localized as well as global conflicts. Through this subversive technique, independent nations can be brought under their control. The three lower boxes show how the elite influence the Military-Industrial Complex. On the left are the defence contractors and their owners – in the middle are the credit creating central banks – and on the right are some of the secretive groups that supply high ranking officials to the MIC. While the subject is outside the scope of this book, Secret Societies lie at the very heart of the globalist movement.

If anyone wants to truly understand the global agenda, they must first grasp the Hegelian dialectic. This is important because many of the conflicts over the last 150 years have been executed this way. They have deliberately allowed socialist ideology to flourish, and if you think it can only happen behind the iron and bamboo curtains, then you are horribly mistaken. When social strategists think about ideological revolution, they are not thinking about a quick change that takes a few weeks or months, they are thinking about a process that takes decades, or in Karl Marx's case, centuries.

Since the Russian revolution of 1917, one-third of the world's land surface has come under Marxist ideology, while the West has been infiltrated by Fabian socialism. What is Fabian socialism? It is a system that creates big government and big corporations, while restricting the rights of the individual, and it is only one surreptitious step away from full socialism.

By understanding that fascism and communism didn't start on the streets of Berlin or Moscow, but with the elite in London and New York, then it's easier to see how those systems might be used as a vehicle for changing the political landscape. While Soviet communism may have officially fallen, the socialist power structure behind it is alive and well, and is being used to powerful effect by the elite.

CHAPTER SIXTEEN

BEYOND THE NATION-STATE

"Whatever shape your world may take in the year 2000 A.D., we can all be fairly sure that it will be one world. Whether through war or through peace, the nations fifty years from now will have learned to enmesh their sovereignties into a single supreme authority. They will have learned to do so because, difficult as it may seem now, no other alternative exists. One world or none at all is the choice."

–Vincent Sheean, Redbook Magazine, 1950

NATIONAL DISORDER

EVEN FROM HIS nineteen-fifties perspective, author, Vincent Sheean knew that something radical had to take place for there to be lasting peace in the world. With the Second World War only five years previous, Sheean, like many others, would have been eager to see an institution like the United Nations make a difference, but as we have seen over the last seventy years, peace is as foreign to us, as it was to them. Interestingly, Sheean might be out on his timing concerning a world government, but I don't think he's out on his prediction.

The dream of establishing a 'One World Government' may be noble in its concept, but because human nature is the way it is, ultimately their dream cannot be brought about by peaceful means. Who in their right mind is going to give up their national sovereignty, as well as their democratic self-determination, so a global body can implement what they think is best for us. So just as the long game will work to a certain degree, in the end, it will be necessary for the elite to bring about world government through a series of wars.

As we have seen, the *financial elite* use war to consolidate wealth and power, and in their eyes, there is no reason why that shouldn't continue. Below are two quotes, one from international banker David Rockefeller, and the other is from Soviet defector Anatoliy Golitsyn, the former KGB Major. Both men give us clues as to what is required for bringing about a 'One World Socialist Government'.

> *"But, the world is now more sophisticated and prepared to march towards a World Government. The supranational sovereignty of an intellectual elite and world bankers is surely preferable to the national auto-determination practiced in past centuries."*
>
> **–David Rockefeller,**
> **Bilderberg Meeting,**
> **1991**

> *"The* [Soviets] *intend... To induce the Americans to adopt their own 'restructuring' and convergence of the Soviet and American systems using to this end the fear of nuclear conflict... Convergence will be accompanied by blood baths and political re-education camps in Western Europe and the United States. The Soviet strategists are counting on an economic depression in the United States and intend to introduce their reformed model of socialism with a human face as an alternative to the American system during the depression."*

–Anatoliy Golitsyn,
KGB Major and Soviet defector,
The Perestroika Deception,
1990

David Rockefeller makes it clear that the elite have no time for national sovereignty, and they believe a 'One World Socialist Government' is inevitable. They are behind the introduction of cross-border trade deals, and multinational corporate takeovers, so they can bring down national independence. But their task of imposing globalisation has not been easy; national sovereignty is still very important to most people, and it will take a lot to get us to relinquish it willingly.

So how are they going to achieve their goal? According to Anatoliy Golitsyn, the communists are going to need assistance from their masters, the *financial elite*, if they are ever going to destroy the West. What that means is, the Soviets are waiting in the wings, until the elite destroy our financial system, and then they will make their move. Interestingly, it has been suggested by some that Political Correctness is Soviet in origin, so could that be the "human face" Anatoliy Golitsyn was alluding to?

If establishing a 'New World Socialist Order' is initiated through the destruction of the US (west) and Russia (east), then it proves they are nothing more than pawns in a game, and it is the elite who are arming both sides with enough firepower to achieve that goal.

That brings us to the military technologies themselves, and why they are ranked in three levels of importance. Most armies have access to the first level, which are known as White World technologies, and include run of the mill declassified equipment. Then there are the Grey World technologies, whose capabilities are only loosely declassified by authorities. They include things like lasers, rail guns, stealth fighters, and nuclear warheads, and because of their restricted availability, they give the globalists a powerful advantage over the smaller nation-states.

The third, and most secretive level, is known as the deep Black World technology programs; they are well hidden inside the intelligence system, and the men who order their use are usually above those of elected officials. Included in this area would be the Aether (scalar) super-weapons, Anti-Gravity, and Bio/Techno-Transhumanism.

So, what about the Aether superweapons, are they being used against the people of the world? According to recent reports, the number and power levels of natural disasters around the planet has increased greatly over the last fifty years. And while the sun is by far the biggest influence over our weather, I also believe there is an artificial component operating.

It is my belief that 'carbon-induced' Global Warming is essentially false, particularly when long-term temperatures are deliberately being falsified by agencies like NASA and NOAA. If the air temperatures are not changing the way they say they are, then why would the elite promote such a lie? I believe there are two main reasons why they do this. The first is, they desire to introduce a global tax system for all citizens of the world, and the 'carbon credit system' under the control of the United Nations is the first step in that endeavour. The second and more sinister reason is, the elite need to cover any tracks the Aether superweapons might leave, for the public are fully aware that something strange is happening to the weather, so it's important the elite use the media to feed us a continuous stream of lies. Remember, we already know about Nikola Tesla's breakthrough in Aetheric-wireless power, and he stated more than once, that a more advanced version of his T.M.T. could control the weather. So, it isn't hard to imagine that somebody out there has developed it into a weapons system, thus giving them the ability to operate a clandestine war against nation-states.

As for the teaching of science itself, I believe the mainstream scientific community is hamstrung because it does not have the Unified Field Theory. How can they say with absolute confidence that they know how the universe works when they still don't truly understand where things like gravity, time, and quantum behaviour come from. Conversely, groups deep within military and political intelligence do possess the Unified Field Theory, and it has given them the scientific foundations for developing equipment that is decades, if not centuries, ahead of the civilian world.

What about the discovery of forbidden technologies by rogue scientists and inventors? It has to be remembered, the elite will stop at nothing to keep their secrets intact. One course of action they take, is to use a piece of legislation called the 'Invention Secrecy Act'. Introduced in 1951, by the United States government, the Act allows certain agencies like the Defence Department to impose restrictions on a patent application if they deem it to be a threat to national security or the national economy. According to the *Federation of American Scientists,* there was 5,680 invention "secrecy orders" in effect as of 2016. While the majority of the secrecy orders have been handed out to government-backed contractors, there have also been orders placed on private inventors, denying them a patent application, and even stopping them from speaking publicly.

That brings us to the man mentioned in *Resonant Power* (chapter 4), who built the device that was able to cancel most of the sound coming out of a Briggs and Stratton engine. Where did his invention go, did it come under the legislation mentioned above? Just imagine his device strapped to the engines of an Apache gunship as it rose above the treeline, all you would hear is the sound of a distant wind before it was all over for you and your terrorist buddies.

Another powerful tool the elite use for suppressing disruptive technologies is the mainstream media. While the mainstream scientific community is deliberately kept in the dark concerning the physical principles of classified technologies, the intelligence apparatus uses the media to discredit anyone who goes against 'their' accepted doctrine. This was seen nowhere more clearly than when two chemists named Martin Fleischmann and

Stanley Pons made their Cold Fusion discovery during the late nineteen-eighties. Even though their experiments into room-temperature fusion have never been properly refuted, the mainstream media went out of their way to diminish the findings and smear their reputations.

Even Nikola Tesla himself complained about the establishment, and how they had painted him to be a poet and a dreamer, which is just a polite way of saying, naive fantasist. Other notable inventors who have stepped into forbidden areas of science are T. Henry Moray (Overunity generator), Royal Raymond Rife (Advanced medical innovations), T. Townsend Brown (Advanced aeronautical propulsion), and Stanley Meyer's (Oxy-hydro combustion engine). They, along with many others have experienced the sinister tactics used to suppress such advancements, and in the case of Stanley Meyers, some believe the inventor was poisoned because he was fully determined to see his revolutionary engine enter the public domain.

Regardless of the advanced science and technologies they possess, as time goes on, it will become harder for the elite to achieve their goals through clandestine means. The reason for that is simple, at some point they will have to come out from behind the shadows so they can fulfil their ultimate plan for mankind.

The one area where they have found it extremely difficult to conceal their presence is when it comes to the dramatic changes taking place in the financial world. In the next segment, we examine the financial agenda, including their plan to destroy the world's currencies, and introduce a global cashless money system.

GLOBAL ORDER

While it's easy to assume there is a conspiracy to overthrow the nation-states of the world, it is another thing to get a consistent thread concerning their intentions. But there is one area which indicates a common link, and that is the colossal mountain of debt accumulating around the world.

It doesn't matter where you look, whether it is individual, corporate, civic or national, everybody is encouraged to operate through credit, even the young are starting their careers with debt from higher learning.

Debt itself is ancient, but the international credit system we have today started around the time of the Second World War. In fact, it was in 1944, that the US dollar became the 'world standard', and the Bretton Woods agreement brought about the World Bank and International Monetary Fund. Since their inception, the WB and IMF have become a powerful credit lending system by which the nations draw upon, and over time many countries have become dependent on their loans. One communique released in 2014 called the *Geneva Report about the World Economy* estimated that global debt had reached a record level of $158.8 trillion. If the interest on that debt was equal to 1%, it would equate to $1.58 trillion per annum, and that's not counting any principal. Yet, in 2018 an article on the *Bloomberg* website states how global debt has now risen to $250 trillion. So, at the international level the world's debt is rising at such a rate that there is no hope of ever paying it back, and because of that, most countries are being pushed to the brink of insolvency.

An article in the *International Business Times* by William White, the former chief economist with the

Bank for International Settlements, states where he thinks this debt mountain will take the world's nations.

> *"Debts have continued to build up over the last eight years and they have reached such levels in every part of the world that they have become a potent cause for mischief... It will become obvious in the next recession that many of these debts will never be serviced or repaid, and this will be uncomfortable for a lot of people who think they own assets that are worth something... The situation is worse than it was in 2007. Our macroeconomic ammunition to fight downturns is essentially all used up," White said. "The only question is whether we are able to look reality in the eye and face what is coming in an orderly fashion, or whether it will be disorderly."*
>
> **–William White,**
> **Ex BIS Chief Economist,**
> **20th January, 2016**

White went on to say how the global commodity market (iron ore, timber etc.), was in oversupply compared to 2007/08. According to him, this would spell bad news for countries like South Africa, Canada, Australia and Brazil, who were not hit so hard by the recession last time. The conclusion that can be drawn from all of this is, the elite have created a giant global Ponzi scheme, where the debt becomes so large the financial system cannot service it any longer, and thus collapses. In this particular case, the Ponzi scheme is where new credit-money is created out of thin air, and pumped into the national economies of the world to keep them buoyant, and then when the *financial elite* are ready to crash the system, they will replace it with a new one. Whether the next crash initiates the global financial changeover remains to be seen, but one thing is for sure, a day of reckoning is coming for the entire world economy.

The *financial elite* are so confident of success that organizations under their control are beginning to reveal the globalist agenda. A quote from an article on the CFR website explains where the world's nations are being led financially, and politically.

> *"There have been several attempts to achieve integration outside of Europe* [EU]—*including the Association of South East Asian Nations (ASEAN), African Union (AU), Gulf Co-operation Council (GCC), and Mercosur in South America,* [The author also includes— North America (NAFTA)—Indian subcontinent (SAARC)—Former Soviet republics (EAEU)—Oceania (CER)—East Asia (EAEC-the proposed economic union for the Far East)]. *There have been innumerable declarations from groupings in Asia, Africa, the Middle East, and South and Central America about the desirability of closer co-operation and even integration... Only after historical reconciliation can countries proceed gradually along the various steps required to create a regional community such as a free trade area, a customs*

union, a single market, a single currency, a common passport area, and a common foreign policy."

**–Council on Foreign Relations,
September, 2010**

As the quote suggests, there is a move to create several (possibly ten) political/economic blocks around the world. This is so nation-states can be wooed into regionally controlled blocks, instead of being forced into an all-encompassing world government. When it comes to the economic component of their plan, it calls for each region to have a single currency like the Euro, as well as full economic integration, and some form of central banking system. It has to be said, while the elite has had issues trying to unify the various regions, it must be remembered, that if they can't encourage globalization through consent, they will force it through conquest.

While the unification of the world's currencies is an important part of the globalist plan, there is another more secretive element mentioned in the Book of Revelations. It is a prophecy by God to the last generation who will see the return of the Messiah, Jesus Christ. In Revelations, Chapter 13, verses 16 to 18 it describes a cashless money system that will be introduced for all peoples of the planet.

*"V16 And he caused all, both small and great,
rich and poor, free and slave, to receive a mark
on their right hand or on their foreheads.
V17 and that no one may buy or sell except*

*one who has the mark or the name of the beast,
or the number of his name.*
   *V18 Here is wisdom. Let him who has under-
standing calculate the number of the beast, for
it is the number of a man. His number is 666."*

–The Book of Revelations

Prior to the current generation, biblical scholars believed the mark would be some kind of tattoo on the right hand or forehead, but since the rise of the information age, and in particular the Silicon Chip, we now know that a lot of information can be held in one of these small devices.

A money system based on that technology would give its operators the ability to send and receive personal information to and from a global mainframe computer. Also, through the GPS satellite system, the chip can be tracked anywhere on the globe, opening the way for the ultimate Orwellian Society.

Some researchers believe the RFID chip used today in tagging animals is the technology that will be used for the mark, while others believe, powerful biochip technology has been secretly developed for this purpose. Whatever the scenario is for introducing the mark is anyone's guess, but the one thing the Bible does make absolutely clear is that we must refuse it at any cost, for your very soul depends on it.

THE KEY

"It is not only the Telegraph that is at fault here", he said. "The past few years have seen a rise of shadowy executives who determine what truths can and what truths can't be conveyed across the mainstream media."

–Peter Oborne, Ex-Chief Political-Commentator for the Telegraph, The Guardian, 2015

MEDIA

FOR THE SAKE of simplicity, it has been necessary to leave certain subjects out of this book, including powerful institutions such as Secret Societies and the Papacy. But there is one particular area that must be given special mention, and that is the Global-Mass-Media. Not because of the financial benefits they bring to the elite, but because of the power they wield over the peoples of the earth, especially now that information can be transmitted to billions of people within a matter of seconds.

If the dream of the *financial elite,* is to establish a 'One World Socialist Government', then they need to indoctrinate us into believing it's the best option for our survival. But can they persuade us into believing the lie that world government is our only hope? If their scheme is to be successful, it will require a lot of propaganda, and it usually starts with a daily diet of negative news.

Like most other areas of our life, the mass media is controlled, and there are three main routes by which this is done. The first is through asset ownership; for example, 90% of all papers, TV and radio stations, and mobile services in the United States, are owned by six giant media corporations. Furthermore, their extensive ownership of many foreign divisions gives them the power to influence people across the world.

Then there are the news collection agencies who on-sell their information to the media outlets. Only three companies dominate this area of journalism. They are Reuters (UK), AP (US) and AFP (France). Collectively they produce twenty-five million words a day, while the next five biggest agencies produce just over a million words daily.

The second form of media control comes from the various off-shoots of the Round Table Groups. Through these secretive organizations, members are placed into policy-making positions within private, as well as state-owned enterprises like the BBC.

Lastly, one of the most effective methods for controlling the media is through the intelligence community. In the West, this covert form of influence was proven to exist when a US government investigation called the 'Church Committee Report' (chaired by Senator Frank

Church D-ID) came out in 1975. From that investigation, it revealed a program called Operation Mockingbird.

In existence since 1948, the program was tasked with recruiting American and foreign journalists into the CIA. Its purpose was for gathering intelligence, but also for projecting CIA policy upon the American and foreign public. Even though much of the report still remains classified, the agency did claim at the time that they were out of the media business.

If their public statement about being out of the media is to be believed, then recent reports must surely bring that declaration into question. According to one article in *MoneyWatch (CBS)*, the Central Intelligence Agency's own technology investment arm, known as In-Q-Tel, invests heavily in tech start-ups. Among other accusations, the article states how In-Q-Tel financed a start-up that later became known as Google Earth, and through affiliate businesses, they are now able to trawl through the databases of Facebook and Twitter.

It is naïve to believe intelligence agencies like the CIA are out of the media industry; if anything the new computer-based forums give them even more power. So if intelligence organisations are influencing the public through the media, then one of the most important tools they use upon the people of the world is the Hegelian method. Through the art of 'divide and conquer' they can impart powerful negative emotions onto the public, which they then use to control both sides of a conflict.

HATRED

"If you tell a lie big enough and keep repeating it, people will eventually come to believe it."

**–Joseph Goebbels, Propaganda Ministry,
German Nazi Party**

Through their control of the mass-media, the elite are able to employ the Hegelian method for dividing various groups within a society, or even across borders. Wars, revolutions and coups d'etat might be the ultimate overt manifestation of hatred, but the covert use of Hegel's method on a peaceful society can be just as effective for keeping people under their control.

By using the media to agitate various groups, a climate of fear can be instilled. If that powerful emotion remains unchecked, it will grow into a deep-seated hatred, and from there culminate in violence, murder and chaos. It must be remembered that the emotions of fear and hatred, lie at the very core of Karl Marx's, Hegelian method.

Also, another important aspect when using the Hegelian method, is to keep everyone's eyes on a patsy, while they manipulate events from behind the scenes. If they can create animosity between races, nationalities, genders, social groups, and religions etc, then their task is made much easier for taking us over. Even people who investigate the global conspiracy have to be extremely careful. It is so easy to fall into the blame game, and point the finger at a particular group. Some say Jews, wealthy Anglo Saxons,

Catholics, or some other group is to blame for the world's ills, and that is where many researchers have made the fatal mistake and fallen into the trap of hatred.

Through years of research into who is at the very pinnacle of this conspiracy, it is my opinion there is a very small group of satanists at the top, and they are fully sold out to their god. These men have no formal religion, no nationality, no social compassion or love, and the most sadistic weapon in their arsenal is the one as old as the serpent himself, and that is they hate anyone who does not think like them.

LOVE IS THE KEY

When hatred gets a hold, it takes control of every decision in a person's life, and there is only one way to break out from under it, and that's by practising love. It is the greatest weapon one can use against fear and hatred, but the individual must choose which emotion they want to live with. In the Book of Proverbs, chapter 10, verse 12, the Bible makes it clear which emotion produces which fruit.

"Hatred stirs up strife, But love covers all sins."

–Book of Proverbs

Through Karl Marx's, Hegelian method, the elite are hard at work instilling hatred and division, so they can generate strife. The Bible states there is only one way to overcome hatred and division and that is by practicing the opposite. In the book of Matthew, chapter 5, verse 44 it states the cure.

> *"But I say to you, love your enemies, bless those who curse you, do good to those who hate you, and pray for those who spitefully use you and persecute you."*

> **—The Book of Matthew**

It is not easy showing love to others, especially when evil is being perpetrated upon us, but by practicing it we tap into God's love and power so that we keep ourselves from becoming full of hate. The commandment is clear, we are to love one another, because in the end love overcomes all evil. In the Book of John, chapter 15, verse 12, it states that love is not just an emotion but it is from God Himself.

> *"This is my commandment, that you love one another as I have loved you."*

> **—The Book of John**

God has given his love to mankind, and it is our job to pass it on to each other, regardless of the cost. But when it comes to traditional western values, we are beginning to see something sinister take place in our midst. With the introduction of Political Correctness over the last few decades, traditional Judeo/Christian love has been turned into a self-serving, humanistic ideology. Furthermore, PC has taken down the Ten Commandments, and promoted the ideology that there are no absolutes, so everything is relative to a person's perception. This has led once conservative western societies around the world into lowering their moral and legal standards, which has led to unprecedented social and judicial upheaval.

As history shows us, when a society replaces good for evil, and evil for good, its demise is not far away. But there is good news, God's intervention in our affairs can change everything, and all we have to do is call on Him in prayer. But to get God to move, we must first acknowledge he exists, and to do that, we must respond to His ultimate gesture of love, His son Jesus Christ. It was his sacrifice on the cross that freed us from our sins, and in the Book of John, chapter 3, verse 16, that love is described in simple terms.

"For God so loved the world that he gave His only begotten son, that whoever believes in Him should not perish but have everlasting life."

–The Book of John

In the end, it all depends on whether we, as individuals, accept the love offering God gave us through his son Jesus, or whether we reject the Biblical statement: that the son of God was sent to die for our sins so we could receive eternal life. God made hell for satan and his rebellious army, but, through deception, the evil one has tricked many people into rejecting God's love, and accepting devilish lies. It must be understood, one of the guiding principles that drives satan and his fallen angels is their hatred towards God, as well as man, whom God created in his image.

What many of those who study the global conspiracy don't understand is, the ultimate battle for global control is spiritual in nature. That makes satan and his army the driving force behind the plan to establish a 'Luciferian One World Order'. When people separate the spiritual from the natural they begin to hate people, and the emotion they never thought they would let themselves come under has taken them over.

The good news is, the devil has lost the war against God, the Bible declares this, but the Book of Revelations states how certain events must take place before the final trumpet is blown, and the Messiah returns. The devil in his own pride, will fight to the bitter end because he knows that when the Messiah returns, his kingdom is gone. That means one thing for the fallen angel the Bible calls the 'serpent of old', and that is he will be one step closer to his final destination, the Lake of Fire.

THE END

GLOSSARY

This glossary contains terms associated with Aether theory, as well as the engineering required to harness it. Other mainstream scientific terms like 'Voltage' use both accepted theory and the Aether interpretation.

Aether
A super-dense field of energy that permeates every point in the universe. The exact opposite to absolute vacuum of space. Various terms have been used throughout history, but the name Aether was commonly used by nineteenth-century scientists.

Aetherist
Someone who believes the Aether model is responsible for unifying all the physical phenomenon in the universe. If correct, Aetherists believe it could be the foundation for the *Unified Field Theory* (UFT).

Aether (scalar) Biodynamics
The use of Remote-Artificial Fields (R.A.Fs) to manipulate or permanently change biological systems, including people.

Aether Density

There are various mathematical estimates, but one thing is agreed upon by most scientists, and that is, empty space, is anything but empty. In fact, every cubic inch is packed with vast quantities of energy. The disagreement between mainstream scientists and Aetherists is, they differ over what that energy is.

Aether (scalar) Geodynamics

The use of Remote-Artificial Fields (R.A.Fs), to manipulate natural phenomena in, on, or above the earth.

Aether (scalar) Psychodynamics

The use of Remote-Artificial Fields (R.A.Fs), to manipulate the psychological state of living organisms, including people.

Aetheric Resonance

Aetheric vibrations at the correct frequency, that allow real physical effects to take place in any system.

Alternating Current

Alternating-Current is an electrical current that flows back and forth along a circuit. The system was chosen over DC because it could be easily transformed into a higher voltage, which allowed power to be sent over long distances.

Capacitor (Electrical)

Unlike a battery, a capacitor is a device that collects and holds electrical energy for a short period of time. Made up of two metal plates and an insulator material (like ceramic) in between, a capacitor is able to change the way electrical current moves along a wire (circuit). In one example, Tesla used capacitors to create strong bursts of D.C. power, which were then sent into his T.M.T. The Serbian also believed the capacitor was the link between electricity and the Aether, just like the magnet and copper winding is the link between magnetism and electricity.

Current

In mainstream electrical theory, current is electric charge flowing down a wire (circuit) and is usually carried by particles called electrons. In Aether theory, current is the result of voltage (Aether) acting upon the individual charges, creating an electric field along the circuit. The latter theory implies the current is not the result of electrons moving down the wire at high speed.

Direct Current

Direct-current is the flow of electrical charges in one direction down a wire. Another commonly used term is Unidirectional.

Dynamic Aether

The concept of the Aether as a highly complex field, which moves energetically in and around the outside of physical bodies like particles and planets. It also moves through 'so-called' empty space itself.

Tesla stated how the Aether is like a gas, and it is common knowledge that gaseous fluids like the atmosphere are anything but static.

Electromagnetic Wave

Sometimes referred to as Transverse or Hertzian waves, EM waves are electric and magnetic fields locked together, and are capable of moving at the speed of light. Modern physics believes electromagnetic waves do not require a medium to travel through. Conversely, Aetherists believe EM waves travel through the Aether like a wave travels on the ocean.

Electrostatic Field

When two objects (or plates) are in close vicinity of each other and have different electrical charge potentials, an electrostatic field manifests between them. If an electrostatic field builds up enough strength, a spark discharge appears between the two objects.

Energy

Energy is the measurement of a systems ability to do work. Just like other forms of energy, Aether is also capable of performing work.

Impulse Current

Sometimes called Tesla, Aether or scalar waves. Impulse currents are not like electromagnetic waves, in that they are not transverse, but are longitudinal in nature (think laser). According to Tesla, they are not bound by the speed of light and also retain their power over great distances.

Impulse Current Frequency

An Impulse Current is made up of a 'chain of pulses' like pearls on a string. The frequency is the number of pulses going past one point every second.

Lines of Force

The all-pervading Aether appears to have 'lines of force'

just like a magnetic field. The 'lines of force' are a representation of the change in pressure, or what some might call, force. Because the Aether is gaseous the 'lines of force' move like a fluid, so they change (expand and contract) in the presence of energy or matter.

Linear Wave

Waves moving in a constant flow over time are considered linear.

Non-Linear Wave

A wave or pressure difference that seems to appear from nowhere. In Aether theory, interlocking stationary waves are capable of creating these effects, with an example being a voltage spike on an electrical circuit.

Photon (Light)

Mainstream science believes light is both a particle and a wave at the same time, and it can move through the vacuum of space without the assistance of a medium.

Conversely, Tesla believed light was a super high-frequency vibration in the Aether. If Tesla was correct, then it must be assumed that Aether creates the resistance as light travels through it, and it's this relationship between the two which creates the speed limit for light waves.

Plasma

Considered the fourth state of matter, it is the most plentiful physical substance in the universe. Modern science proposes that plasma is just an electrified gas. But according to Aetherists, the all-pervading field can also be seen in some plasma phenomena. For example, plasma is observed in its electrical state as Fork lightning, while in its Aetheric form it sometimes appears as Ball lightning or Stress lights.

Potentials

There are different types of *potential* energy, but essentially it means the unrealised ability to perform work. For example, water in a glass contains *potential* energy, then when it is poured out it is realised energy.

In the case of electricity, it flows from a higher *potential* to a lower one, and it will only stop when there is equalization.

In the case of Aharonov and Bohm's 1959 scientific paper that involved *Potentials*, they are described as a non-physical realm which directs particles.

But according to Aetherists, those *Potentials* are believed to be physical in nature and are capable of causing real effects. Also, it is proposed that they are a part of the all-pervading Aether, just like a wave is a part of the ocean.

Radiation

Nuclear radiation is defined as energy particles or rays that are given off from a radioactive element (such as uranium) as it decays. Tesla, on the other hand, believed it was the result of inflowing Aether being converted into radiation.

Remote-Artificial Field. (R.A.F)

A 'Remote-Artificial Field' (R.A.F), is a manmade Aether field. When two or more longitudinal Aether currents (scalar waves) are sent out from the transmitters, they are made to cross at a specific point on the globe. The R.A.F. can be customized into various shapes and power levels for whatever purpose the operator requires. Also, it can be placed at any point in, on or above the earth.

Remote = The effect is created at a distant site.

Artificial = It is manmade.

Field = It is sent out as Aether energy, but when it reaches the target site (Interference Zone) it becomes a field of EM energy.

Remote-Tuning (Artificial Resonance)

It is where Tesla would cause an object at any distance to come under resonance. When the correct frequency was obtained in the object, a variety of effects would manifest from what appeared to be nowhere.

Squeezing (pinching) – lines of force

In the presence of matter, the Aetheric-'lines of force' begin to squeeze (contract) together. This creates mechanical pressure on matter, which causes real effects to take place.

Stationary (Standing) Waves

A wave that is sent out from point A to point B, and then back to Point A, just like an *echo*. If the wave is sent out at the correct frequency, the troughs and peaks will line up 180° out of phase thus creating a stationary wave.

Static Aether

The concept that the Aether is a non-moving field. This is analogous to physical bodies like atoms and planets moving through the all-pervading field like a submarine moves through a static ocean.

Supercharging

In Tesla's early experiments he discovered a mysterious force moving down the wire when the high voltage separated from the electric current.

Superluminal

Any speed above the speed of light. In Aether theory, impulse waves are able to move from zero to infinity, allowing for instantaneous communications across the length of the universe.

Tesla Coil

Based largely on the works of Sir Oliver Lodge, the *Tesla Coil* is designed to release its EM electrical charges, which sometimes manifest as arcs (lightning bolts).

T.M.T. Tesla-Magnifying-Transformer

Otherwise known as an *Electrical Oscillator*, or *Tesla Magnifying Transmitter*, the T.M.T. might look similar to the modern Tesla Coil, but it does the exact opposite. While the Tesla Coil is designed to produce amazing electrical phenomena, the T.M.T. is designed to retard those effects inside the coils so the voltage can flow freely. That is why it must be insulated from external electrical effects, otherwise, inbound charges will create explosive effects.

Virtual-Field

The Virtual-Field is the entire Aetheric field inside, as well as around the atom. It includes the *Core* (nucleus), the outer layers called the *Foundations* (electron shells), and the *Corona* (the area surrounding the atom). It also includes the currents, stationary waves, and resonance that creates their internal structures. NOTE: Subatomic particles are also believed to be made from Aether.

Virtual-Field – Core

The *Core* of the Virtual-Field is the structure inside an atom's nucleus. Made up of a toroidal current, as well as a perpendicular North/South current through the centre. According to Aetherists, the Virtual-Core is the invisible part of the atomic nucleus. NOTE: The North/South current also extends out past the Foundations, and into the Corona.

Virtual-Field – Foundations

The area where the atoms electron shells reside.

Virtual-Field – Corona

The area of Aether space surrounding the atom, or in quantum physics, the area known as the Van der Waals zone.

Voltage

Also called the electromotive force, mainstream electrical science states that voltage is the pressure (or potential) placed upon moving electrons and their charges. In Aether theory, voltage is Aether trapped on the circuit, causing the charges (not electrons) to move at close to the speed of light down a circuit.

Wattage

A form of measurement. It is a unit of power that is produced or consumed over a certain period by an electrical appliance or circuit.

Made in the USA
Middletown, DE
24 February 2025

71763360R00184